XENOPUS: The South African Clawed Frog

XENOPUS: The South African Clawed Frog

E. M. Deuchar
Department of Anatomy,
University of Bristol

A Wiley—Interscience Publication

John Wiley & Sons
LONDON · NEW YORK · SYDNEY · TORONTO

Library of Congress Cataloging in Publication Data:

Deuchar, Elizabeth Marion.
Xenopus: the South African clawed frog.

"A Wiley-Interscience publication."
Bibliography: p.
1. Xenopus. 2. Xenopus laevis. 3. Embryology—Amphibians.
4. Development biology. I. Title.

QL668.E265D48   597'.8   73–18927
ISBN 0 471 20962 7

Printed in Great Britain by
William Clowes & Sons, Limited
London, Beccles and Colchester.

Preface

This book arises out of some five years' work, since it includes subject-matter that I started collecting together in 1969 for a Biological Review entitled '*Xenopus laevis* and Developmental Biology' (Deuchar, 1972). I have now added much more material to this and have broadened the discussion of some of the earlier topics, as well as including some more recent publications. The first three chapters of this book give, besides a fairly extensive survey of the anatomy of *Xenopus*, brief descriptions of the species and subspecies and their geographical locations, and some of the more recent as well as the earlier work on the physiology of this animal. The rest of the book still reflects my predominant interest in developmental biology, however, and is almost entirely concerned with work on the embryonic and larval development of *Xenopus laevis*. This has been a favourite amphibian for laboratory work for some 35 years now, because it is easy to keep and responds well to hormone injections. Being aquatic throughout its life, it does not require any elaborate cage or terrarium, such as may be needed for some tropical frogs: it can simply be kept in a tank of water all the year round. Coming from regions not far from the Equator, it has a more adaptable breeding season than frogs of temperate countries, and so responds to treatment with gonadotrophic hormones at any time of year. Moreover, it produces very large numbers of eggs: several hundred at one spawning and not the mere two or three per day that are achieved by some newts and salamanders.

I was first introduced to *Xenopus* in Edinburgh in 1949 when I was trying to do a Ph.D. project in experimental embryology and the breeding season for newts was over. I shall always be grateful to the late Dr. Cecil Gordon for this useful introduction: also to Dr. Bruce Hobson of the Pregnancy Diagnosis Laboratory in Edinburgh, whose interest in the responses of both male and female frogs to gonadotrophins provided me with large numbers of *Xenopus* embryos (many of which—alas!—perished in the cold and vibration as I bicycled back with them to the Institute of Animal Genetics, over the cobbled streets and tramlines which still existed then!). Since then, *Xenopus* has been a standby for all my work in amphibian embryology, in London and in

Bristol. Successive batches of these robust frogs have withstood the rigours of air flights from Cape Town to Britain, long waits at airports in the fog, and overcrowded conditions on arrival. I often spare a thought for the hundreds of others travelling to research laboratories in Europe and the United States each year. The work carried out on them has taken a leading place in many fields of recent biological interest, as I have tried to show in the following pages.

During the writing of this book I have often been conscious of omitting to mention some authors and of doing insufficient justice to others. I may unwittingly have misrepresented some authors' findings, too. To all these people, I apologize, and I hope that they will write and tell me of any such failings. Many friends and colleagues have offered help, and I am very grateful to them: especially the experts in South Africa whom I met on my visit to collect material for the review, in 1970. At the University of the Witwatersrand, I had useful discussions with Professor B. I. Balinsky, Dr. John Balinsky, Dr. Vivian Gabie and Dr. Rosemary Caunter, and at Cape Town University Dr. Naomi Millard gave me some invaluable reprints of her work. It was a privilege also to meet Professor H. Zwarenstein there and to receive encouragement and the loan of a rare monograph from him. There were many others who offered hospitality and transport: for instance, Dr. Toerien at Stellenbosch took me on a 'Xenopus-hunt' to the Jonkershoek Reserve and the surrounding country.

In the preparation of the final manuscript of this book, I have been most grateful to Dr. M. J. Manning for reading it all through and offering some helpful suggestions and further references for the immunology section of Chapter 10. Dr. R. C. Tinsley has also given much kind help and criticism on Chapter 1. I should like, too, to thank the Editor of John Wiley and Sons for his continual guidance and for arranging professional help with some of the illustrations.

Several authors have kindly allowed me to reproduce illustrations from their papers: I am most grateful to them, and also to their publishers for copyright permissions. These permissions are acknowledged on p. vii.

Finally, I owe thanks and apologies to my husband, family, colleagues and students who have borne a certain amount of neglect during the last stages of my writing of this book. Like some of them, I have often wanted to quote *Ecclesiastes 12*.12:—'of making many books there is no end; and much study is a weariness of the flesh'.

Why *does* one write any book? Because, I suppose, the subject (and in this case, the animal) is there, and is fascinating, and one wants to share it with others while one's enthusiasm lasts. I hope to have been able to pass on some of my enthusiasm for the development biology of *Xenopus* to readers of this book.

Bristol, September 1973   ELIZABETH M. DEUCHAR

Acknowledgements of Copyright Permissions

Permission has kindly been granted by the following authors and publishers, for reproductions that appear in the halftone and line figures listed below. These permissions are gratefully acknowledged.

Halftones

Figures 1.4, 1.5 and 1.6: by permission of Professor A. W. Blackler and the publishers of *Revue Suisse de Zoologie.*

Figure 1.7 (a): by permission of Professor A. W. Blackler and Academic Press, Inc.

Figure 1.7 (b): by permission of Dr. R. C. Tinsley and the Zoological Society of London.

Figures 4.5, 4.6, 4.7, 4.8 and 4.9: by permission of Dr. H. P. Stanley and Academic Press, Inc.

Figure 4.10: by permission of Professor B. I. Balinsky and the editors of *Acta Embryologiae Experimentalis.*

Figure 5.3: by permission of Dr. S. W. de Laat and Academic Press, Inc.

Figures 6.1, 6.2 and 6.3: by permission of Dr. M. R. Kalt and the Company of Biologists Ltd.

Figure 10.7: by permission of Dr. R. Hauser and Springer-Verlag.

Linework

Figure 3.5: by permission of Dr. J. C. Poynton and E. J. Brill, Leiden.

Figure 4.16: by permission of Dr. D. D. Brown and Academic Press, Inc.

Figure 5.13: by permission of Dr. R. Weber and the publishers of *Revue Suisse de Zoologie.*

Figure 7.1: by permission of Dr. P. H. Tuft and the S.E.B. Symposium Secretary.

Figure 7.2: by permission of Dr. T. E. Schroeder and the Company of Biologists Ltd.

Figure 8.1: by permission of North-Holland Publishing Co.

Figure 8.2 (a): by permission of Professor C. H. Waddington and *Nature.*

Figure 9.1: by permission of Drs. H. Woodland and J. Gurdon, and the Company of Biologists Ltd.

Figure 9.3: by permission of Dr. Y. W. Kunz and *Experientia* (Birkhaeuser Verlag).

Figure 10.2: by permission of Dr. M. Jacobson and Academic Press Inc.

Figure 10.6: by permission of the publishers of *Revue Suisse de Zoologie.*

I am particularly grateful to Dr. D. Brown, Dr. M. Stanisstreet, Dr. A. W. Blackler, Dr. R. Tinsley, Drs. S. C. Reed and H. P. Stanley, Prof. B. I. Balinsky, Dr. S. W. de Laat, Dr. M. Kalt and Dr. R. Hauser for providing me with original prints of their photographs.

Contents

1

The Discovery of *Xenopus* in its Natural Habitats

The genus of African frogs which we now know as *Xenopus* has been given a variety of names in the past since its first discovery by Daudin (1803). Daudin called it 'Le Crapaud Lisse' and suggested the Latin name *Bufo laevis* for this smooth-skinned, toad-like creature. The Latin word 'laevis' means 'smooth', and Daudin evidently thought that his new animal belonged to the most familiar genus of toads, *Bufo*. He did note also, however, that it had some resemblances to *Pipa*, a genus of aquatic frogs found in South America and popularly named the 'Surinam toad'. Subsequent studies of the anatomy of *Xenopus* have shown that it does in fact belong to the frog family Pipidae. So, although *Xenopus* is usually referred to as 'The South African Clawed *Toad*', it is really problematical whether it should be called a toad or a frog.

The great French comparative anatomist, Cuvier, was the first to give a fairly detailed description of *Xenopus* (Cuvier, 1829), and he proposed the name 'Dactylethra' (meaning 'finger sheath') on account of its sheath-like claws. This name continued to be used until the 1890's. Mayer in 1835 gave a further description of 'Dactylethra', with illustrations which leave no doubt that this was identical to *Xenopus*. There were at first differing views about a specific name to go with Dactylethra: Tschudi in 1838 suggested *Dactylethra borei*, and Dumeril and Bibron in 1841 called it 'Dactylethra of Cuvier'. It is not clear whether all these early authors were referring to the same species of *Xenopus*, for it was not until 1858 that Günther pointed out that more than one species existed: one in the Cape area and in West Africa which he called *Dactylethra laevis*, and one with a more northerly distribution, described by Peters in 1844 as *Dactylethra muelleri*. *D. muelleri* differed from *D. laevis* in having a prominent tentacle under the eye and a spur at the base of the first toe on the hind limb. Gray in 1864 also recognized these two species, but proposed the name *Dactylethra capensis* (Cuvier) for the form found in the

Cape region. This name would however have been unjust to Daudin, who was the original discoverer of the animal.

The generic name *Xenopus*, meaning 'peculiar foot', was first suggested by Wagler (1827) in a footnote to a paper by Boie, but was not adopted until much later when Leslie (1890) described living *Xenopus* for the first time, in their natural habitat, instead of merely preserved specimens as hitherto. Leslie's description was, surprisingly, published in the Proceedings of the Zoological Society of London and not in a South African journal. He had found that the 'Plathanda', as it was called by the colonists at Port Elizabeth, was very common in local rivers and pools there. He noted that it ate fish, insect larvae and even occasionally its own tadpoles. He observed also the presence of an internal valve on the nostrils, characteristic of air-breathing aquatic animals. The 'ticking' call of the male was, he thought, produced by friction of the glottis against the borders of the Eustachian opening from the middle ear, as the air was drawn into the buccal cavity from the lungs.

The first description of the spawning and the tadpoles of *Xenopus* was by Beddard (1894) and was based on six specimens in the London Zoo which had been imported from Cape Town. By the time that Bles (1905) wrote his comprehensive account of the breeding and the development of the embryos and larvae, the official name of the species from the Cape was *Xenopus laevis* (Daudin), as it is now. Its popular name in South Africa is still 'Plathander', or 'Platanna', or even 'Platie' for short. Its external appearance and habits are described in most books on African amphibians. Since it is used for dissection in biology courses at schools and universities in South Africa, its anatomy has also been described simply (see: *Dissection of the Spiny Dogfish and the Platanna*, Millard & Robinson, 1955). We shall leave consideration of the anatomy to the next chapter, and will now review points of interest about the habits, external features and geographical distribution of the various species of *Xenopus*.

1. External Features, Habits and Life History

Compared with most other common frogs, *Xenopus* looks flattened dorsoventrally (hence the nickname 'Platie') and its body is large relative to the size of the head (see Figure 1.1). Its very long hind legs, with long, webbed toes, are splayed out sideways instead of being flexed under the body as in terrestrial frogs adopting a crouch position. The much shorter forelegs are also extended laterally, and play no significant part in supporting the body when it is resting on a surface. This flattened shape and posture are clearly more suitable for swimming than for progress on land, and *Xenopus* is an entirely aquatic animal, unlike most other common frogs and toads. It cannot survive for long on dry land, but when occasionally *Xenopus* does have to cross land surfaces, it is capable of moving reasonably fast by a series

Figure 1.1 Male and female *Xenopus laevis* in amplexus during spawning in the laboratory. Viewed from above, $\frac{9}{10}$ life size. The eggs are seen in the background, scattered over the bottom of the aquarium

of lolloping leaps, landing flat on its belly each time and slithering off course if the surface is too smooth. The skin is smooth and slimy, the slime having an unpleasant musky smell which may deter predators (including man!), but not being toxic except perhaps to other amphibians.

The head of *Xenopus* is flattened like the body, and the eyes are smaller and less prominent than those of terrestrial frogs. The dorsal colouring of both head and body in *Xenopus laevis* is a mottled greenish-grey when the animal is on a neutral-coloured background, but it can change to darker or to paler on either dark or light backgrounds (see Chapter 3). The ventral surface is yellowish-white in *Xenopus laevis*, but has more vivid colouring in other species. Males and females are similar in coloration, but the female is larger than the male and has cloacal papillae (see Figure 1.2).

The adult *Xenopus* is a carnivore, feeding fairly indiscriminately on all kinds of living or dead animal material that may be present in the stagnant pools in which it lives. It is very rarely found in running water (despite Leslie's report, see above). When feeding, the forelimbs are used to push food into the mouth after it has been torn up by the claws which are present on the innermost toes of each hind foot (hence the name 'clawed toad'). It has been reported to eat its own tadpoles when other food is short, and in the laboratory will sometimes eat its own eggs. The tadpoles feed on detritus in the water or on the bottom, by means of a filter-feeding mechanism (see Chapter 2).

A striking feature of *Xenopus*'s behaviour is its regular need to come to the surface of the water and gulp air. If the water is sufficiently shallow, it will stand for some time on its toes with its nostrils just above the surface, apparently inhaling air. As we shall see later, *Xenopus* probably depends on air breathing more than do other frogs, since it does not have a very efficient means of respiration through the skin.

An aquatic habit is not without its risks in the warmer areas of Southern Africa. When some of the pools dry up in summer, *Xenopus* has to burrow into the mud to avoid desiccation, and here it remains quiescent, aestivating, until the next rainy season. Breeding usually takes place during a period of three to five months per year, when the pools are full: this is in spring, in temperate regions. The male when courting emits a grating croak and grasps the female round her pelvic region, remaining in this position of amplexus (see Figure 1.1) while she lays her eggs. The male has dark, sticky hairs on the digits and forearms (see Figure 1.3) which help him to maintain his grip on the female, and he emits sperm at intervals, which have to travel forward to reach the eggs as these are extruded. Normally from 500 to 1,000 eggs are laid at one spawning, over a period of up to 24 hours. The eggs have individual jelly capsules (cf. Figure 1.1) like those of newts and salamanders, and are not enclosed in any continuous jelly mass like the spawn of other frogs. The larvae hatch within 3 days after spawning, and reach metamorphosis in about 2 months if food is plentiful and they are not overcrowded. Just after

Figure 1.2. The sexual characters. Female, slightly less than life size, viewed from dorsal surface. Note the projecting cloacal papilla at her hind end

metamorphosis the young froglets are very delicate and vulnerable, and may occasionally develop rickets under laboratory conditions unless the · diet is optimal or is supplemented with bone meal and cod-liver oil. But if they survive these early hazards, they can grow to adulthood and start breeding in about eighteen months.

There are surprising stories of the longevity of *Xenopus* in captivity, but it is not certain how long they live in the wild. Hewitt (quoted by Wager 1965) says that some specimens of *Xenopus* survived for fifteen years in laboratories in Cambridge. In the wild, very large females are sometimes caught, which are probably ten or more years old, as the females grow continuously throughout

Figure 1.3. The sexual characters *(continued)*. Head-on view of male in breeding condition, showing blackening of the forearms and inner surfaces of the digits. (Linear magnification × 2)

life. At 12–13 years old in the laboratory, females have been found to weigh 100–150 gm. Males, however, are fully grown at 2–4 years, so their size is no clue to their age after this. One symptom of ageing is a decrease of elasticity in tendon collagen, as in mammals (Brocas & Verzar, 1961). Malignant tumours may also appear in *Xenopus*, and a number of these have been described (Elkan, 1960). Some types are transmissible after periods in culture (Balls & Ruben, 1964, 1967). Like many fish and amphibians, *Xenopus* may develop fungal skin infections. 'Red-leg' is also a common infection, thought to be caused by the bacillus *Aeromonas* (Tinsley, personal communication 1973). It may be cured by adding a few drops of a 1 % solution of 'mercurochrome' (di-brom-oxy-mercuri-fluorescein) to the water for several days.

Xenopus individuals in the wild harbour many parasites in the gut, liver, blood and other organs. Several of these parasites were described some time ago by Weinbrenn (1925), and recently Tinsley (1973) has embarked on a detailed comparison of the parasites found in different species of *Xenopus*. They include protozoans, spirochaetes, nematodes, trematodes, mites and leeches. The cercarian larva of one species of trematode attacks the lateral-line organs, making them become melanosed and thus easier to see (Elkan & Murray, 1951). The normal complement of parasites causes no ill-health to the frogs, but in adverse conditions, starvation, etc., they may succumb to fungal or bacterial infections.

The foregoing remarks apply chiefly to the commonest species, *Xenopus laevis*, since this has received most study. The various species and subspecies of *Xenopus*, besides harbouring different parasites, differ in their external features, their geographical distribution and habitats, and in their mating calls, as we shall now describe in the following section.

2. Species and Subspecies of *Xenopus*

The most recent survey of the differences between the several species included in the genus *Xenopus* (Wagler) is that by Tinsley (1973), and the following account incorporates some of his findings. In the past, eight separate species have been described, but two of these, '*Xenopus calcaratus*' (Peters) and '*Xenopus poweri*' (Hewitt) have since been reclassified. '*X. calcaratus*' has been found to be identical to *X. tropicalis* (described below) and '*X. poweri*' has been shown to be a subspecies of *X. laevis*. So, with the discovery of one new species in Uganda by Tinsley (1973), there are now seven recognized species of *Xenopus*. In addition, seven subspecies are now known within the species *X. laevis*. Some of the salient features which distinguish all these forms are given below, and are also summarized in Table 1.1 below. External features of some of the animals are shown in Figures 1.4–1.7.

Xenopus laevis (Daudin) (Figure 1.4) is the original form described as 'Bufo laevis' by Daudin in 1803, and is the commonest and widest-ranging species. It is found throughout the temperate regions of South and West Africa, and extends as far north as Malawi, Rhodesia and Mozambique also, though it is less common here. Rarely, it has been found in the Kruger National Park, in one locality near Pretoriuskop. Now that other subspecies have been described, this form should be designated *X. laevis laevis*. As noted earlier (p. 4) it has a mottled greenish-grey coloration dorsally, and cream underparts with no characteristic markings. Its hindlimbs are massive and its forelimbs small, and there are claws on its three innermost hind digits. Its head has a rounded jaw margin, not pointed as in some other *Xenopus* species, and the eyes are medium-sized, with very small subocular tentacles (see Figure 1.8). The mating call of the male is a grating sound of medium pitch, rather like a squeaky water-tap being turned on.

Xenopus laevis victorianus (Ahl) (Figure 1.5) was discovered by Ahl (1924) and found again in the eastern Congo by Curry-Lindahl (1956), both authors regarding it as a separate species. It has since been designated a subspecies of *X. laevis*, however: this was confirmed by Blackler, Fischberg and Newth (1965), who showed that it could interbreed with *X. l. laevis*. It is smaller than *X. l. laevis* and has an olive-green dorsal colouring with scarcely

Figures 1.4–1.7
Some species and subspecies of *Xenopus*. All are shown life size (approx.)

Figure 1.4. *Xenopus l. laevis*. (a) Female, dorsal view. (b) Male

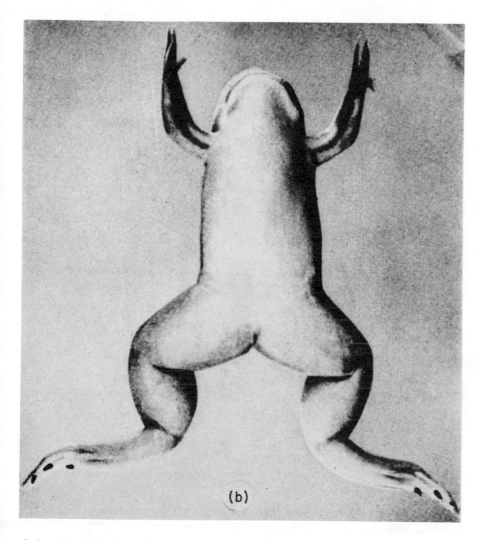

(b)

entral view. From Blackler, Fischberg and Newth, 1965

Figure 1.5. *Xenopus l. victorianus.* (a) Female, dorsal view. (b) Female, ventral view. From Blackler, Fischberg and Newth, 1965

Figure 1.6. *Xenopus l. petersi.* (a) Female, dorsal view, (b) Female, ventral view. From Blackler and Fischberg, 1968

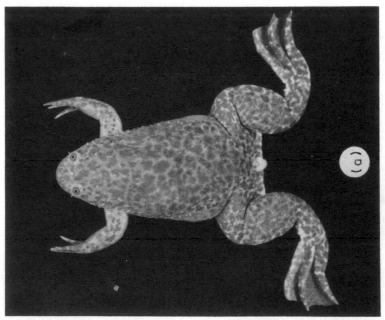

Figure 1.7. (a) *Xenopus muelleri*: female, dorsal view. From Blackler and Gecking, 1972. (b) *Xenopus vestitus*: dorsal view. From Tinsley, 1973

any of the dark mottlings seen in *X. l. laevis*. On the under surface it shows orange-yellow tints on the thighs, unlike the whitish-yellow of *X. l. laevis*. The eggs of *X. l. victorianus* are smaller and paler than those of *X. l. laevis*, but embryonic development is alike in the two subspecies, apart from minor features such as the time of appearance of the chromatophores and their distribution. Hybrids between *X. l. laevis* and *X. l. victorianus* are intermediate in size and show a dorsal colour pattern with aggregates of spots on a dark green background while the ventral surface is partly spotted, unlike either subspecies.

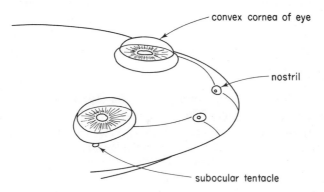

Figure 1.8. Diagram of externals of head region in *Xenopus laevis*, showing the very small subocular tentacle

Xenopus laevis poweri (Hewitt) was discovered by Hewitt near the Victoria Falls in 1927, and he thought it to be a separate species. Both Schmidt and Inger (1959) and Poynton (1964) have concluded that it is a subspecies of *X. laevis*, however. It has a northerly distribution, from Angola, Zambia and Tanzania down to the borders of South and South-West Africa. This subspecies differs from *X. l. laevis* in having a darker dorsal colour, a yellower ventral surface, larger eyes and narrower nostrils. Its colouring is not so dark as that of *X. l. petersi*, described below.

Xenopus laevis petersi (Bocage) (Figure 1.6 above) is another form which was originally thought to be a separate species (Bocage, 1895), but Blackler and Fischberg (1968) showed that it will interbreed with *X. l. laevis*, so must be a related subspecies. They describe it as smaller than *X. l. laevis*, averaging 70–74 mm body length. It has elongated black patches on the dorsal surface and a bright orange ground colour ventrally, with violet-black patches on it. The eggs are smaller than those of *X. l. laevis* and have a cream-coloured marginal zone. Hybrids have the same dorsal colour as *X. l. petersi*, but are paler ventrally, with smaller spots. *X. l. petersi* is found in Angola and in the northern regions of Rhodesia.

Xenopus laevis bunyoniensis is a form which was found by Loveridge (1932) in the lakes of Uganda, and it has been re-described by Tinsley (1973) from museum specimens. There has been some dispute as to whether this is a separate subspecies: Parker (1932) regarded it as identical to *X. l. laevis* and thought that Loveridge had come across some animals infected with the larvae of a fluke, causing the lower eyelids to swell and the eyes to bulge! Blackler and Fischberg (1968) thought that it could be the same as *X. l. victorianus*. After measuring many of its external features and comparing it with other species, Tinsley (1973) felt that many more specimens of *X. l. bunyoniensis* needed to be studied before it could be decided whether it should be regarded as a subspecies or not. He was not able to study the coloration in fresh specimens, but Loveridge (1932) and Parker (1936b) agree in describing this frog, which is smaller than *X. l. laevis*, as having a bright orange ground-colour ventrally with dark spots, and a very dark dorsal coloration.

Xenopus laevis borealis was found by Parker (1936b) in collections from eastern Africa—i.e. Somalia, Ethiopia and Kenya. It is distinguished from *X. l. laevis* by having a projecting metatarsal tubercle on the hind foot, and by having a larger size range: its body length is 100 mm or more.

Xenopus laevis sudanensis (Perret) was found on the plateau of Adamaona and the plain of Bénoué in the Cameroons by Perret (1966), and is described by Tinsley (1973) as having nostrils of similar width to those of *X. l. laevis*. Perret noted that its dorsal coloration was olive green, with either small black spots or larger black blotches, while the ventral was yellow-brown, sometimes with orange tints. It has a short, broad prehallux which distinguishes it from *X. l. laevis*, *X. l. petersi* and *X. l. borealis*. It is smaller than *X. l. laevis* but larger than *X. l. bunyoniensis*—the maximum size of the females being 65 mm. It has a longer tibia than *X. l. victorianus*.

We now pass on to the different species of the genus *Xenopus*.

Xenopus muelleri (Peters) (Figure 1.7(a) above) has for a long time been recognized as a species distinct from *X. laevis*. It was first described as 'Dactylethra muelleri' by Peters in 1844. It has a more northerly distribution than *X. laevis* and is found in N. Zululand, the Kruger Park, the lowlands of Transvaal, Rhodesia and Zambia, and also in Malawi, Tanzania, Kenya, the Congo, and savanna areas of southern and western Africa (Arnoult & Lamotte, 1968). Its distinguishing points are its smaller size, more pointed head, less massive hind limbs and longer forefingers than *X. laevis*. It also has a much longer subocular tentacle than *X. laevis*, and the underside of the thighs is usually dark yellow in adults, or orange in young frogs. The belly may be spotted, blotched or dark in colour. *X. muelleri* is said to have a distinctive mating call, 'tut-tut', like a pencil tapping on glass, and unlike the grating, long-drawn-out rasp of *X. laevis*.

Xenopus tropicalis (Gray) and *X. fraseri* (Boulenger) have been re-described

Table 1.1. Species and subspecies of *Xenopus* recognized at present

Name	Distribution	Distinguishing external features
Xenopus laevis:		
(a) *X. l. laevis*	Temperate regions of S. and W. Africa: also Malawi, Rhodesia and Mozambique. Rarely at Pretoriuskop.	Size range 70–110 mm (females). Green-grey dorsal, cream ventral. Rounded jaw margin. Very small subocular tentacle. Grating mating call.
(b) *X. l. victorianus*	Eastern Congo.	Smaller size range than *X. l. laevis*. Olive green dorsally: orange-yellow ventral thighs. Eggs smaller and paler than those of *X. l. laevis*.
(c) *X. l. poweri*	From Angola, Zambia and Tanzania down to borders of S. and S.W. Africa.	Darker dorsal than *X. l. laevis*. Yellow ventrally. Larger eyes and narrower nostrils than in *X. l. laevis*.
(d) *X. l. petersi*	Angola and Rhodesia.	Size range 70–74 mm. Elongated black patches dorsally: bright orange with violet-black spots ventrally. Eggs smaller than those of *X. l. laevis*.
(e) *X. l. bunyoniensis*	Uganda lakes.	Smaller size range than *X. l. laevis*. Very dark dorsally: orange with dark spots ventrally.
(f) *X. l. borealis*	E. Africa (Somalia, Ethiopia and Kenya).	Larger than *X. l. laevis* and with metatarsal tubercle.
(g) *X. l. sudanensis*	Cameroons.	Size range between *X. l. laevis* and *X. l. bunyoniensis*. Olive green dorsal with black spots: yellow-brown ventral, sometimes with orange tints. Short, broad prehallux.
Xenopus muelleri	Wide distribution further north than *X. laevis*. (N. Zululand, Transvaal, Rhodesia, Zambia, Malawi, Tanzania, Kenya, Congo and W. Africa).	Smaller than *X. laevis*. Long subocular tentacle. More pointed head, less massive hind limbs than *X. laevis*. Underparts spotted and thighs yellow or orange. Mating call a sharp 'tut-tut'.
Xenopus tropicalis and *Xenopus fraseri*	Equatorial forests of W. Africa: Gabon, Nigeria and Cameroons.	Ratios of body/limb lengths and eye/eyelid sizes. *X. fraseri* has larger lower eyelid than *X. tropicalis*.

Table 1.1 (contd.)

Name	Distribution	Distinguishing external features
Xenopus gilli	Cape and Stellenbosch, becoming rare.	Large blotches dorsally: yellow with dark spots ventrally. Pointed head. No subocular tentacles. Long urostyle. Mating call 'Chee-chee'.
Xenopus clivii	N.E. Africa (Somalia, Ethiopia).	Irregular spottings dorsally. Long subocular tentacle. Body and limb proportions like *Xenopus muelleri*.
Xenopus vestitus	Kigezi, Uganda, Eastern Congo and other areas in Central Africa.	Size range small: 45–55 mm. Head pointed. Limbs shorter than in *X. laevis*. Golden brown dorsally with dark lines on head. Cream with yellow/orange patterns ventrally.

by Parker (1936a) and classified as separate species. Both are confined mainly to the equatorial, forested regions of West Africa (i.e. Gabon, Nigeria and the Cameroons), but have been reported also at Fernando Po and in East Africa. Tinsley (1973) has observed considerable variation in the external features within each species, but it is possible to distinguish the two species on the basis of ratios of various body measurements: for instance, the relative sizes of the eye and of its lower lid, and of limb length to body length. On this basis, there are closer resemblances between *X. fraseri* and *X. laevis* than between *X. fraseri* and *X. tropicalis*. Other distinguishing features of these two species are given by Parker (*loc. cit.*). He points out that in *X. fraseri* the lower eyelid extends over the lower half of the eye, and there are several elongated pustules on the head as well as a long subocular tentacle (Figure 1.9). *X. tropicalis* has only a very small vestige of the lower eyelid, in the anterior corner of the eye, and it has a shorter subocular tentacle (Figure 1.9) and smaller, more numerous pustules on the head than *X. fraseri*. Colour comparisons between the two species are not helpful, since Sanderson (1936) found two quite different colour types in *X. tropicalis* from the Cameroons. Those living in forests were brown dorsally and yellowish-grey beneath, while specimens found in a river were olive green dorsally and white beneath.

    Xenopus gilli (Rose and Hewitt) was described by these authors in 1927. It is confined to the Cape and Stellenbosch areas, and is now becoming rare owing to the gradual drainage of the swamp areas of the Cape Flats and encroachment of buildings onto them. *X. gilli* is much smaller than *X. laevis*, averaging 50 mm in length. It has a few large, dark blotches on its dorsal surface, including a prominent pair on the head. The underparts are yellow with black spots: hence Rose (1962) calls it the 'sago tummy toad'. It has a more pointed head than *X. laevis*, smaller eyes, a flatter abdomen and a longer dorsal urostyle in the pelvic girdle. It has no subocular tentacles. The mating call is described by Rose as a loud and strident 'chee-chee'. Rose and Hewitt regarded *X. gilli* as more closely related to *X. tropicalis* than to *X. laevis*, since its pectoral girdle is less arciferous than in *X. laevis* (cf. Chapter 2) and its metatarsals are less elongated than in *X. laevis* or *X. muelleri*.

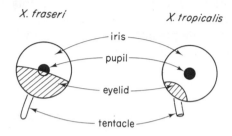

Figure 1.9. Eyelids and subocular tentacles compared in *Xenopus fraseri* and *Xenopus tropicalis*. Redrawn after Parker, 1936

Xenopus clivii (Peracca) is a species found in north-east Africa (Somaliland and Ethiopia) and is not very well known. Peracca (1898) obtained his specimens in Eritrea. Tinsley describes the morphology of the head and proportionate size of the limbs as being close to those of *X. muelleri* and *X. l. laevis*. The dorsal surface has irregular spottings. There is a long subocular tentacle and the eyes are medium-sized: these and other features noted by Tinsley make it similar in appearance to *X. fraseri*.

Xenopus vestitus was discovered recently by Laurent (1972) and described in detail by Tinsley (1973, 1974). Tinsley at first named it *X. kigesiensis*. He found specimens in Lake Mutanda in south-west Kigezi, Uganda, where it seems to have supplanted the form described by Loveridge as *X. l. bunyoniensis* (see above, p. 14). *X. vestitus* is a relatively small species, the largest males being 45 mm and females 55 mm long. The head is pointed, the eyes are small and a subocular tentacle is present. The limbs and digits are shorter than those of *X. l. laevis* The coloration of this new species is golden-brown dorsally, with no spots. The head may be lighter in colour, but has a dark line between the eyes and a dark collar round the occipital region. The ventral parts are cream, with yellow or orange patterns on them. Figure 1.7(b) above shows one of Tinsley's photographs of this species.

Tinsley (1973) pointed out that the several species of *Xenopus* fall into more than one dichotomous grouping. For instance, if one regards the claws as diagnostic of relationships, then *X. tropicalis*, *X. fraseri* and *X. clivii* form one group, having a fourth horny claw on the outer hind toe, whereas in *X. laevis*, *X. muelleri*, *X. gilli* and *X. vestitus* this fourth 'claw' is only a fleshy protuberance. Alternatively, grouping them according to limb proportions, *X. tropicalis*, *X. gilli* and *X. vestitus* are the short-limbed forms, in contrast to *X. laevis*, *X. muelleri*, *X. clivii* and *X. fraseri*. When eye sizes are compared, *X. tropicalis* and *X. vestitus* are seen to have smaller eyes than the other five species. Since none of these groupings coincide, it is doubtful if they throw any light on how closely related the individual species are. Only artificial hybridization experiments, with a careful follow-up of the degree to which normal embryonic development was possible in the hybrids and the state of their nuclei, might possibly indicate the degrees of genetic difference between the various species. Colour patterns within species and in hybrids are more consistent than tints, which vary with environment or even with diet. (Frogs fed on *Tubifex*, a worm containing haemoglobin, may show pinkish ventral tints). Tymowska and Fischberg (1973) have noted differences in chromosome number in some species and subspecies of *Xenopus*. Most have 36 chromosomes, but one new variety which they name *X. Ruwenzoriensis* is apparently polyploid, with 108 chromosomes.

3. Relationships of *Xenopus* to the Rest of the Amphibia

As we shall see in Chapter 2, one of the features of *Xenopus* which has intrigued comparative anatomists for many years is that it has several resemblances to the urodeles (tailed amphibians: newts and salamanders) although it clearly belongs to the Order Anura (tailless amphibians: frogs and toads). *Xenopus* also has features which show that it belongs to the Pipidae, as we noted earlier. The Pipidae are characterized by having no tongue and no movable upper or lower eyelid. Their other characteristics are mostly skeletal: these are listed in Table 1.2. Together with the family Discoglossidae, the Pipidae constitute a suborder of the Anura, called Opisthocoela because the centra of their vertebrae have concave articulating surfaces posteriorly. Two other skeletal features of the suborder are that the scapulae are smaller than in other suborders and that the ribs are free, for at least part of their development, instead of being ankylosed as in other anurans.

The classification within the suborder Opisthocoela has been set out by Noble (1931) and the relevant points are summarized in Table 1.2. Noble divides the Pipidae into two subfamilies, Pipinae and Xenopinae. Within the Xenopinae three genera are known: *Xenopus*, *Hymenochirus* and *Pseudohymenochirus*. Of these three, *Hymenochirus* is the closest to *Pipa*, and *Xenopus* the least like *Pipa*, with *Pseudohymenochirus* in an intermediate position. *Hymenochirus* is found in tropical Africa and differs from *Xenopus* in having webbing on the forefingers as well as on the hind digits. It has claws on the inner hind digits, but no teeth.

In his descriptions of *Xenopus*, Noble lists certain characters that may be

Table 1.2. Placing of *Xenopus* within the Suborder *Opisthocoela*. [After Noble (1931)]

Order: *ANURA (SALIENTIA)*.

Suborder Opisthocoela:

Vertebrae opisthocoelous with well-fused centra.
Scapulae shorter than in other anuran suborders.
Free ribs, in either larva or adult.

Includes two families: Discoglossidae and Pipidae.

Family Pipidae:

Aquatic, aglossal frogs.
No movable eyelids except in *Pseudohymenochirus*.
Ribs free in larva but ankylose at metamorphosis.
Pectoral girdle not always firmisternal.

Includes two sub-families: Pipinae and Xenopinae.

Subfamily Xenopinae:

Simple, pointed digits.
Three inner toes have horny claws.
Tadpole has long tentacles and right and left spiracles.

Includes three genera: *Xenopus*
        Hymenochirus
        Pseudohymenochirus.

---

regarded as 'primitive' and others that are 'specialized'. But as we shall see in the following chapter, there is considerable doubt still as to which features of *Xenopus* are specializations acquired secondarily for aquatic life, and which represent the absence of characters suited to life on land. Were the ancestors of *Xenopus* once terrestrial, and has this genus subsequently returned to an aquatic habit? Or, has *Xenopus* not yet evolved the ability to live on land? These are still unresolved questions. What is indisputable, however, is its undoubted relationship to the Pipidae, a group of frogs, and not to the Bufonid toads. The misnomer 'toad' has probably stuck to *Xenopus* because of its apparent ugliness compared with many common frogs. But if it is seen in all stages of its development, in clean conditions and in adequate space so that it can demonstrate its agility in swimming, *Xenopus* is a by no means unattractive animal. The larvae are beautifully streamlined and transparent, and the long-limbed froglets just after metamorphosis have the same charm as many other juvenile animals. To offset the gross bulk and greedy habits of the adult, it has so many interesting anatomical, physiological and developmental features that even it could be said to have endeared itself to experimenters and become an attractive laboratory animal.

In the following chapters we shall be considering examples of the very wide range of experimental work that has been done on *Xenopus*, and the many interesting findings that have emerged.

2

The Anatomy of *Xenopus*

Just as the external form of *Xenopus*, with its flattened body and its limbs splayed out sideways, is very different from that of most other frogs and toads, there are also several features of its internal anatomy that are unusual. Because of these peculiarities, some of which seem to be primitive, others to be specializations for aquatic life and others to be suggestive of a relationship to the urodeles, *Xenopus* has long been an object of interested study by comparative anatomists. In this chapter we shall mention first some of the striking features which originally aroused most interest and controversy, and will then go on to a more systematic description of the internal anatomy, compared with that of the common British frog, *Rana temporaria*. A few points of comparison with the Urodele *Salamandra*, whose anatomy has been very fully described by Francis (1934), will be noted too.

1. Special and Controversial Features of the Anatomy

The earliest detailed account of the anatomy of both adults and larvae of *Xenopus* was that by Dreyer (1913, 1914). He noted some apparently primitive features in this animal, as well as others which he thought indicated a relationship to the urodeles rather than to any other anurans. But as we saw in the last chapter, Noble (1931) later identified in *Xenopus* several characteristics of the Pipidae, and *Xenopus* has since been assigned to this family in the Anuran suborder Opisthocoela (see Table 1.2 above). Noble considered *Xenopus* as the most primitive of the three genera: *Xenopus*, *Hymenochirus* and *Pseudohymenochirus*. He listed as primitive features in *Xenopus* the following: (i) the pectoral girdle is not 'firmisternal' (i.e. the sternum is not ossified, and the coracoids are not fused to it); (ii) there are separate prevomers in the bones of the skull; (iii) the pterygoid bones are not fused round the opening of the Eustachian tube as they are in *Hymenochirus*. Noble regarded some other features of the skull in *Xenopus* as specialized, however: the reduction of the maxillary bones, the reduction and forward

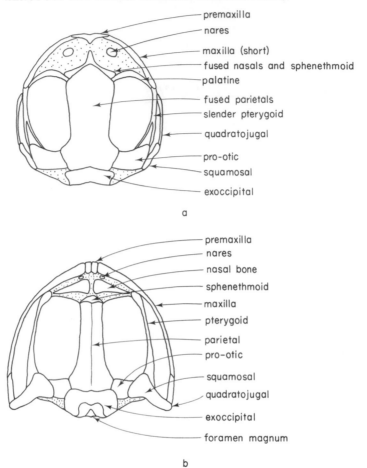

- premaxilla
- nares
- maxilla (short)
- fused nasals and sphenethmoid
- palatine
- fused parietals
- slender pterygoid
- quadratojugal
- pro-otic
- squamosal
- exoccipital

a

- premaxilla
- nares
- nasal bone
- sphenethmoid
- maxilla
- pterygoid
- parietal
- pro–otic
- squamosal
- quadratojugal
- exoccipital
- foramen magnum

b

Figure 2.1. Dorsal views of skulls of *Xenopus* and *Rana*, compared.
(a) *Xenopus*. (b) *Rana*. Linear magnification × 2. Dotted areas are
cartilage

extension of the squamosals, and the fusion of the sphenethmoids and
parasphenoids to encase the brain. Some of these points show in Figure 2.1,
in which dorsal views of the skulls of *Xenopus* and *Rana* are compared. The
greater thickness of the dorsal skull bones in *Xenopus*, although not remarked
on by Noble, is most evident when one is attempting to remove these in order
to dissect out the brain.

To those of us who are not comparative anatomists it may seem arbitrary
to designate minute, apparently trivial variations of skull structure as either
'primitive' or 'specialized'. But of course it is *only* the skeletal features
that can be compared with fossil forms as well as with species that still exist.

Because of the little fossil material available, the choice of features to compare may be very limited, and the criteria by which they are judged as primitive or specialized may be very unconvincing to an outsider. In the last analysis, experts may designate as primitive those features which show resemblance to animals either living or extinct which have been regarded on other grounds as being primitive. They may then regard as specialized any features which are not seen in more than a very few other species, and/or seem to be suitable adaptations for the particular mode of life of the animal.

Figure 2.2. Head of *Rana*, side view, to show its more prominent eyes than in *Xenopus* and the presence of an external ear-drum (cf. Figure 1.8.)

In *Xenopus*, it is fairly easy to pick out some features of its anatomy that are well suited to aquatic life, though it is not easy to decide if these are adaptations acquired secondarily in evolution, or are really primitive, 'pre-terrestrial' characters. For instance, the eyes in *Xenopus* are smaller and less prominent than those of *Rana* (cf. Figure 1.1 and 2.2) and their field of vision is directed upwards rather than forwards. This is advantageous to an animal which spends much time lying on the bottom of pools and whose predators and prey probably approach it from above. *Xenopus* also has a highly convex cornea which stands up well above the surface of the head (cf. Figure 1.3 and Figure 1.8) and is evidently a specialization to improve the range of vision overhead. If one next considers the organs of hearing, *Xenopus* differs from terrestrial forms like *Rana* in having no visible ear-drum externally (cf. Figures 1.8 and 2.2). This might be regarded as a primitive feature, since the ear-drum is an organ specialized for receiving delicate, airborne vibrations and is not found in fully aquatic vertebrates. However, on removal of the skin behind the eye, one finds in *Xenopus* a flat disc of cartilage on the outer wall of the otic capsule. No such disc is present in *Rana*. The cartilaginous disc in *Xenopus* could function in the same way as the ear-drum of *Rana*, but because of its greater robustness, could analyse better the vibrations carried through water. These vibrations are also detected by the lateral-line organs, which in *Xenopus*, unlike other Anura, persist in the adult and are not lost at metamorphosis. There is in fact an elaborate lateral-line system with both dorsal and ventral components, visible externally in *Xenopus* (cf. Figure 2.3). It provides characteristic markings on the otherwise smooth skin, which lacks the knobbly protrusions that are seen in *Rana*. Witschi (1955) observed a further specialization in *Xenopus* for the detection of vibrations: the 'bronchial columella'. He described this as a delicate rod of fibrocartilage which transmits vibrations from the round window of the inner ear to the lung of that side. One of its functions is thought to be to transmit pressure changes in the environment to the lungs, and to stimulate a regulation of the volume of air in the lungs, so that the animal can adjust the depth to which it is immersed in the water.

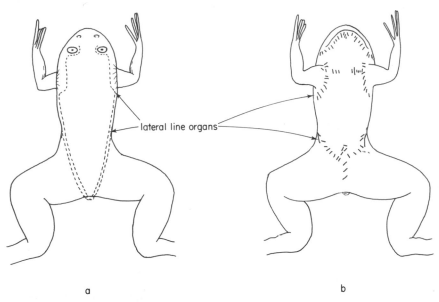

lateral line organs

a   b

Figure 2.3. Dorsal and ventral views of lateral-line organs in a female *Xenopus*. (a) Dorsal. (b) Ventral

Another interesting and apparently adaptive feature in *Xenopus* is that the lungs are larger in proportion to its body size than are the lungs of other anurans. They are about four times as long as the lungs of *Rana*, for instance (cf. Figures 2.15 and 2.16). They develop very early too: soon after the larva has hatched. It is thought that their earliest function may be, as indicated above, to regulate the depth at which the larvae are immersed. It is also probable that *Xenopus* depends more on its lungs for respiration than do other amphibians, in both larval and adult stages. The relatively sparse blood supply to the skin and the gills (cf. section 7 below) suggests that in *Xenopus* neither of these is a very efficient respiratory organ.

2. The Anatomy of the Skeleton: Special Features

There are many unusual features in the limb skeleton and limb girdles in *Xenopus* which are easy to relate to its predominantly aquatic life, in which swimming is the normal mode of progression. The short, slim forelegs (Figures 1.1–1.4) are not used much either for locomotion or for support of the body, but they carry out rapid criss-cross movements to shovel food into the mouth during feeding. The pectoral girdle therefore needs to be flexible rather than robust, and in fact it is not ossified to the same extent as in other anurans. Figure 2.4 compares the pectoral girdles of *Xenopus* and *Rana*. As noted earlier, there is no rigid 'firmisternal' attachment of clavicles or coracoids to the midline in *Xenopus*. Both the coracoids ventrally and the scapulae dorsally have thin, cartilaginous medial borders which overlap one another in the mid line as the limbs move. The overlapping articulation of the coracoids is known as 'arciferous' (Devilliers, 1924; Van Pletzen, 1953) and

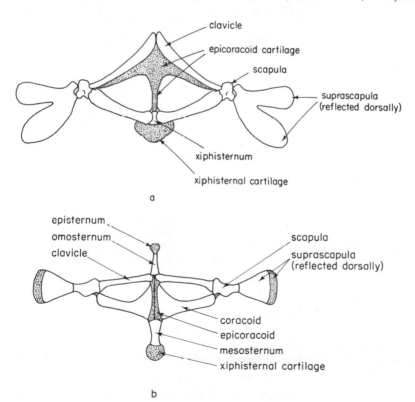

Figure 2.4. Pectoral girdles of *Xenopus* and *Rana*, compared: ventral views. (a) *Xenopus*. (b) *Rana*. Linear magnification ×2. Dotted areas are cartilage. Scapulae shown as if in same plane.

is regarded as characteristic of more primitive groups of amphibians. The pelvic girdle of *Xenopus*, by contrast, is particularly robust, with the pubes ossified instead of being cartilaginous as in other anurans (cf. Figures 2.5 (a) and (b)). It is also closely articulated with long, wing-like extensions of

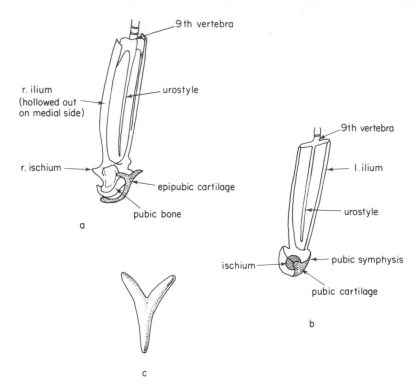

Figure 2.5. Pelvic girdles and urostyles of *Xenopus* and *Rana*, and the ypsiloid cartilage of *Salamandra*, for comparison. (a) *Xenopus*. (b) *Rana*. (c) Ypsiloid of *Salamandra*, redrawn after Francis, 1934. A right ventrolateral view is shown in (a) and (b)

the transverse processes of the 9th vertebra (see Figure 2.6), and the internal faces of the ilia are concave for this articulation. Another special feature is the presence of a long epipubic cartilage in the mid-ventral line: this is also shown in the Figure. There is no epipubic cartilage in other anurans, and it is thought by Hoffmann (1930) to be homologous to the ypsiloid cartilage of urodeles (Figure 2.5 (c)). The epipubis develops from a pair of cartilaginous strips on each inner margin of the rectus abdominis muscle (Devilliers, 1925). There is still controversy as to whether or not its presence in *Xenopus* indicates a relationship to the urodeles.

The presacral vertebral column of *Xenopus* is shorter than in other anurans, perhaps giving it greater strength and rigidity in association with the large and strong pelvic girdle. The development of the vertebrae has been described by Mookerjee and Das (1939) and Smit (1953). They regard some features as primitive compared with other anurans: for instance, the retention of a band of hyaline cartilage derived from the notochordal sheath, and the

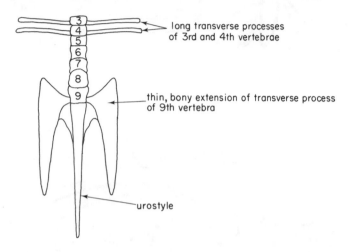

long transverse processes
of 3rd and 4th vertebrae

thin, bony extension of transverse process
of 9th vertebra

urostyle

Figure 2.6. Ventral view of vertebral column of *Xenopus* (semi-diagrammatic). The first two vertebrae are not shown. Approx. × 2

'opisthocoelous' condition of the vertebrae (i.e. with concave articulating surfaces caudally). Opisthocoelous vertebrae are seen in all members of the suborder Opisthocoela, and are regarded as primitive (Goodrich, 1938). However, examination of the vertebral column of adult *Xenopus* leaves no doubt that it has specializations too, giving it extra strength. There are long, robust transverse process on the 3rd and 4th vertebrae (Figure 2.6), besides the striking wing-like extensions on the 9th vertebra which articulate with the pelvic girdle.

3. The Nervous System

Considerable attention was paid by earlier anatomists to the nerves of the head region in *Xenopus*. Paterson (1939a, b) observed some resemblances to urodeles in the layout of these cranial nerves. For instance, there is no maxillary branch of the Vth cranial nerve, and instead its profundus branch innervates the maxillary region. In Figure 2.7 the cranial nerves of *Xenopus* and *Rana* are compared. Dreyer (1914) followed their development in

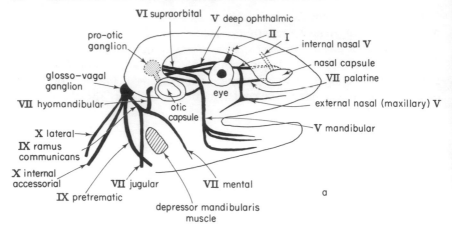

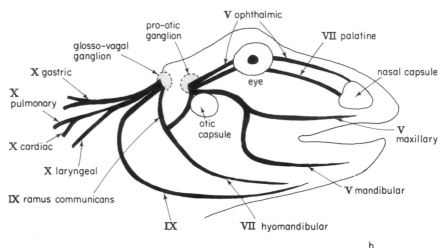

Figure 2.7. Major cranial nerves of *Xenopus* and *Rana*, compared: lateral views (diagrammatic). (a) *Xenopus*. Based partly on Millard and Robinson, 1955. (b) *Rana*. Based partly on Whitehouse and Grove, 1947

Synopsis of the main nerves in Xenopus (from Millard and Robinson):

V. Trigeminal. (a) deep ophthalmic, from prootic ganglion, crosses orbit and branches into: (i) internal nasal supplying skin of nose.

(ii) external nasal supplying skin of upper jaw and lateral nasal region.

(b) mandibular, from prootic ganglion, runs out between eye and auditory capsule and supplies lower jaw region.

VII. Facial. (a) supraorbital branch: runs close to mandibular V and divides into superior ophthalmic and infraorbital branches at the eye. Supplies skin of eye region, and lateral-line organs here.

(b) palatine—from prootic ganglion, passes to roof of buccal cavity. Lies behind eye.

Xenopus, and recorded as a primitive feature the fact that there is originally only a single ganglion for nerves IV, V, VI and VII, and also only one for nerves IX and X. In the adult, too, it can be seen that the branches of nerves V and VII still emerge from a common, pro-otic ganglion and that there is a common glossopharyngeo-vagus ganglion for nerves IX and X. The same is true for *Rana*, however, so this is not a condition peculiar to *Xenopus*, and cannot be regarded as 'primitive' compared with other anurans.

The spinal nerves in *Xenopus* have an essentially similar layout to those of *Rana* (see Figure 2.8), but they have been numbered differently in the two genera by some authors. In both *Xenopus* and *Rana* the most anterior pair of spinal nerves seen in the adult is the hypoglossal, but this is strictly speaking the second of the original spinal series, as no. 1 is lost during development. So the hypoglossal nerve is called no. 2 in Millard & Robinson's description of *Xenopus* (1955), though it is called no. 1 in *Rana* by Whitehouse and Grove (1947). Hence all the succeeding numbers are different in the two accounts, though it can be seen that there are equivalent pairs of nerves in the two genera. Nos. 3 and 4 (or '2' and '3' in *Rana*) are stout, and branch to form the brachial plexus, then nos. 5, 6 and 7 ('4', '5' and '6' in *Rana*) are very slender nerves running to the body wall. In *Xenopus* nos. 6 and 7 run alongside each other for part of their length, encased in a common sheath of connective tissue. Nerves 8, 9 and 10 ('7', '8' and '9' in *Rana*) form the lumbar plexus and are very stout, in contrast to the fine 11th ('10th') nerve which is difficult to find, especially on the right side according to my experience.

(c) hyomandibular—from prootic ganglion, passes through auditory capsule and divides into: (i) jugular branch, to interhyoid muscles: visceral motor fibres.
     (ii) mental branch, out to mandibular depression, skin and lateral line organs.

IX. Glossopharyngeal. From glosso-vagal ganglion, 2 branches: (a) ramus communicans—joins hyomandibular, and goes to skin. (b) pretrematic—to walls of pharynx and mouth. Viscero-sensory fibres. The post-trematic branch is present in the larva but lost in the adult.

X. Vagus. From glosso-vagal ganglion—2 branches.
(a) lateral—to body sense organs.
(b) intestino-accessorial—to skin, muscles of shoulder, hyoid, larynx, gut, lungs and heart.

Other cranial nerves: I Olfactory—supplies nasal organ.
       II Optic—to eye.
       III, IV and VI. Oculomotor, Pathetic and Abducent—to eye muscles.
       VIII Auditory—to ear.

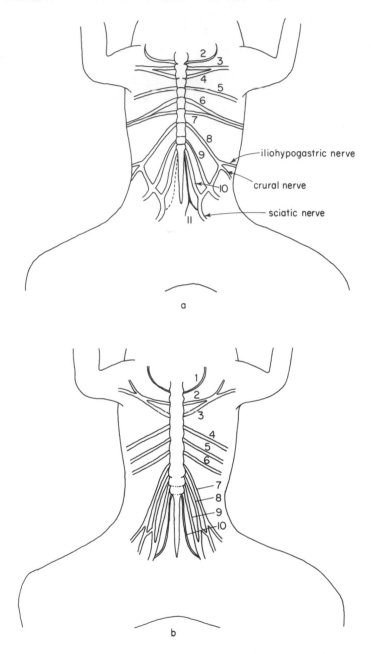

Figure 2.8. Dissections of the spinal nerves of *Xenopus* and *Rana*, compared: ventral views, after removal of the viscera. (a) *Xenopus*, (b) *Rana*. 11 is dotted on R.H.S. of (a) because it was not found

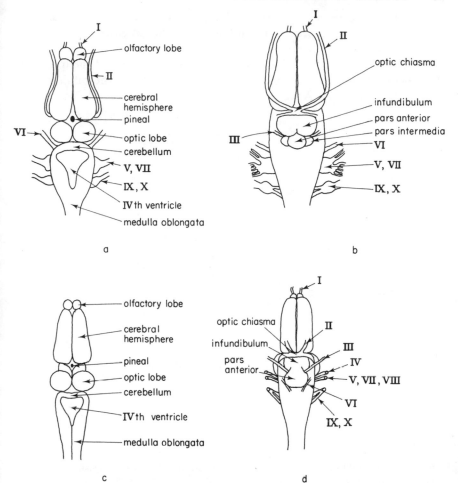

Figure 2.9. Brains of *Xenopus* and *Rana*, compared from dissected specimens. (a) *Xenopus*, dorsal view. (b) *Xenopus*, ventral view. (c) *Rana*, dorsal view. (d) *Rana*, ventral view. Linear magnification × 5

The central nervous system of *Xenopus* shows no major morphological differences from other anurans. External views of the brains of *Xenopus* and *Rana* are compared in Figure 2.9. It will be seen that there are only minor differences in shape: the cerebral hemispheres are a little shorter and broader in *Xenopus* than in *Rana*, and the optic lobes slightly smaller. The pituitary gland of *Xenopus* has prominent lateral projections of the pars intermedia, showing on either side of the pars anterior (Figure 2.10). In *Rana*, on the other hand, the pars intermedia is not evident from the ventral aspect. Rimer (1931) described the pars tuberalis, which forms part of this intermediate region, as being particularly large in *Xenopus*, and trilobed. She also described a process extending from the anterior lobe into the pituitary stalk and very much resembling a structure in urodeles which is regarded as a homologue of the *tuber cinereum*, a portion of the forebrain floor between the hypothalamus and the neural lobe of the pituitary (cf. Figure 2.10 (a), (b)). Atwell (1941) regarded the process in *Xenopus* as

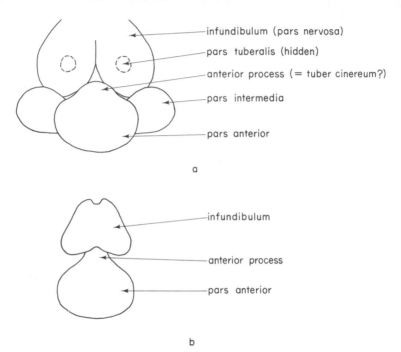

Figure 2.10. Detail of external appearance of pituitary gland of *Xenopus*, ventral view (redrawn after Rimer, 1931), compared with that of *Salamandra* (redrawn from Francis, 1934). (a) *Xenopus*. (b) *Salamandra*

simply part of the anterior lobe, however, and described the pars tuberalis as having only two lobes. He agreed with Rimer, nevertheless, that there were general resemblances to the pituitary of urodeles (cf. Figure 2.10 (b)). More will be said about the functions of the pituitary of *Xenopus* and its internal anatomy, in Chapter 10.

4. Pineal and Thyroid Glands

Rimer (1931) also described the pineal and thyroid glands of *Xenopus*. The pineal is larger, more branched and more glandular in its histological structure than is the pineal of other anurans studied, and it bears a closer resemblance to the pineal of urodeles. There is no parapineal, and the paraphysis is small. The external position of the gland is seen in Figure 2.9.

The thyroid of *Xenopus* was also described by Rimer (1931) and she noted that it frequently showed accessory follicles. These have been found to take up radioactive iodine, which suggests that they may function as normal thyroid tissue and secrete thyroxine. There are also parathyroid glands, which lie further posterior than those of *Rana*, according to Shapiro (1933) (see Figure 2.11) and whose histology is more like that of urodelan than of other anuran parathyroids. There are sometimes three pairs of parathyroids in *Xenopus*, instead of the two pairs typical of anurans.

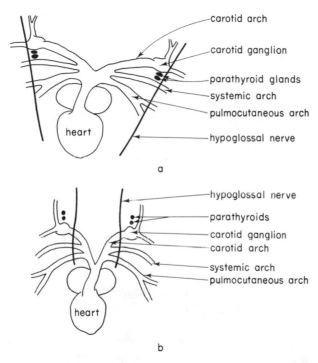

Figure 2.11. Positions of parathyroid glands in relation to main arterial trunks in *Xenopus* and *Rana*, compared. Redrawn after Shapiro, 1933. Ventral view diagrams. (a) *Xenopus*, (b) *Rana*

5. Muscles

Grobbelaar (1935) and Jones (1938) made detailed studies of the musculature in *Xenopus*, paying special attention to the limb muscles and their lack of specialization for life on land. Jones noted that there was no rotation in the radio-ulna of the forelimb, and that both the pollex (first

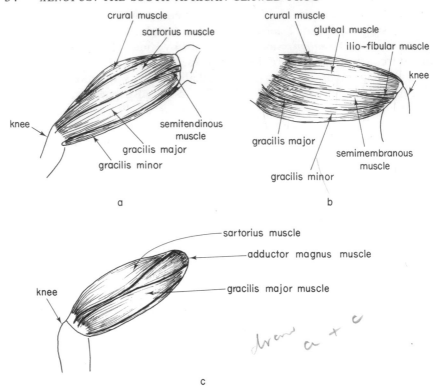

Figure 2.12. Thigh muscles of *Xenopus* and *Rana*, compared. (a) *Xenopus*, ventral view (b) *Xenopus*, dorsal view. (c) *Rana*, ventral view

digit) and the pre-pollex (an extra digit lateral to the first) were absent. He also noted the absence of adductor and abductor muscles capable of drawing the limbs under the body into a position suitable for jumping on land. In the leg, the semitendinous muscle replaces the adductor ventrally, and the semimembranous muscle replaces the abductor dorsally. These, as their names suggest, have relatively few muscle fibres and much membranous or tendinous tissue. The thigh muscles of *Xenopus* and *Rana* are compared in Figure 2.12. In the foot, *Xenopus* has stronger and more widely radiating digital muscles than *Rana* and other terrestrial anurans (cf. Figure 2.13): these are used in the strong swimming movements.

There are minor general differences in the trunk musculature of *Xenopus* and *Rana*, which are most readily seen when the skin is deflected prior to dissection. In a fresh specimen of *Xenopus*, the muscles are covered by a shiny mucous layer which makes their sheaths iridescent and the boundaries of the individual muscles somewhat unclear. The general texture of the

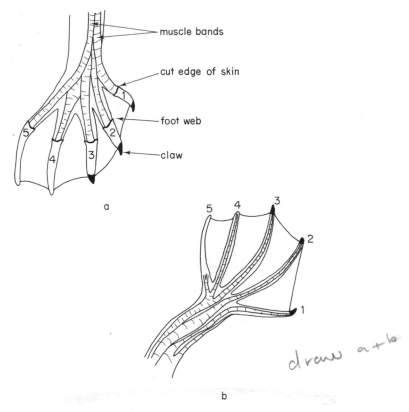

muscle bands

cut edge of skin

foot web

claw

draw a + b

Figure 2.13. Simplified drawings of the large digital muscles in the foot of *Xenopus*. (a) Dorsal view. (b) Ventral view. Skin has been dissected off as far as distal phalanges. Muscles are cross-hatched. Digits numbered according to usual conventions. Linear magnification × $1\frac{1}{5}$

muscles is smoother and flimsier in *Xenopus* than in *Rana*. In *Rana*, the muscle bundles are well defined and the fibres are yellower and tougher than those of *Xenopus*. Both the rectus abdominis and the pectoralis muscles are thicker and more prominent in *Rana* than they are in *Xenopus*. Ventral views of the trunk muscles are compared in Figure 2.14.

Dorsally, *Xenopus* has two abdominal muscle bands which have aroused interest among comparative anatomists. One thin muscle band on each side runs forwards from the ilium bone of the pelvic girdle, then splits into two components: a dorsal one which spreads over the dorsal surface of the oesophagus and over the roots of the lungs, and a ventral aponeurosis of connective tissue which covers the liver and is continuous with the pericardium. Between them, these two components partially separate the lungs

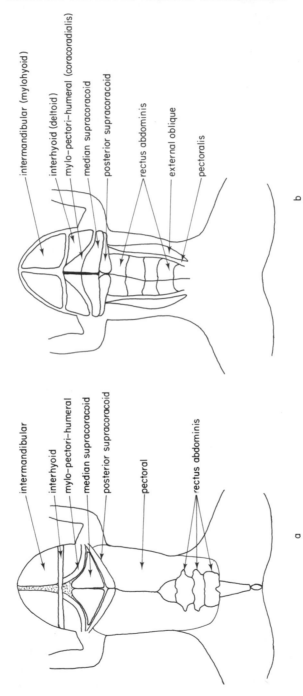

Figure 2.14. Outlines of muscles of trunk in *Xenopus* and *Rana*, viewed from ventral surface after removal of skin. (a) *Xenopus*. (b) *Rana*

from the peritoneal cavity, and the dorsal band could exert compression on the lungs by contraction of its muscles. There are relatively few muscle fibres in it, however, and it extends over very little of the lung's surface (cf. Figures 2.15 (a) and (b)). Keith (1905) regarded these dorsal muscle bands, which are found in other anurans as aponeuroses, as homologues of part of the mammalian diaphragm. *Xenopus* is unique in having some muscle fibres present as well as connective tissue here, and this may reflect its greater dependence on pulmonary respiration than other anurans.

6. General Morphology of the Viscera

A general dissection of the viscera in *Xenopus* reveals several further differences from the terrestrial frog *Rana*, which has for a long time been used as an amphibian 'type' specimen in zoology teaching in Britain. Some of the differences have already been listed by Leadley-Brown (1970), and I have stressed here only the features which seemed most striking in my own

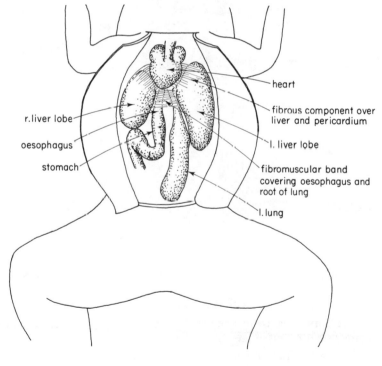

Figure 2.15. Muscle bands attached to lungs in *Xenopus*. (a) Ventral view of dissected specimen: l. lung only shown, and stomach deflected to r. (b) Lateral view, from Keith (1905)

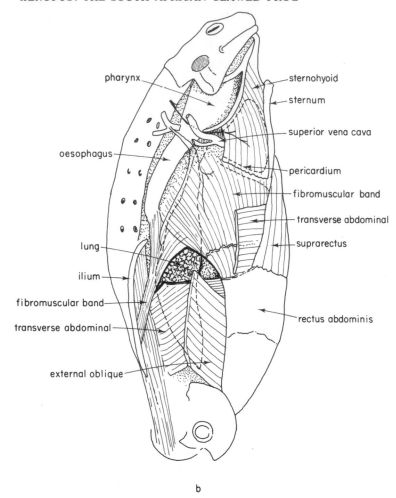

b

Figure 2.15. (continued)

dissections of *Xenopus* and *Rana*. They are illustrated where possible, in Figures 2.16 ff.

First, on ligaturing the anterior abdominal vein before opening the abdominal cavity in a fresh specimen, one notices the flimsiness of the midline musculature overlying this vein in *Xenopus*. It can easily be peeled off the vein, unlike in *Rana* where the ligature has to be tied round muscles and vein together. Next, as soon as further incisions are made into the body wall, the length and large capacity of the lungs in *Xenopus* become evident, for these protrude at once through the incisions. The lungs extend right to the posterior end of the abdominal cavity in *Xenopus*, whereas in *Rana* they

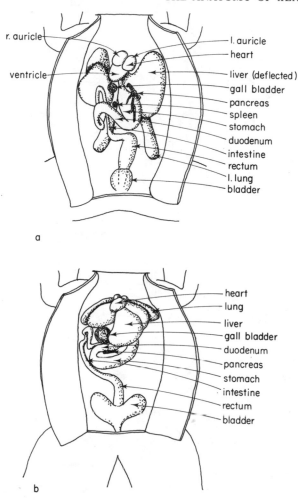

r. auricle

ventricle

l. auricle

heart

liver (deflected)

gall bladder

pancreas

spleen

stomach

duodenum

intestine

rectum

l. lung

bladder

a

heart

lung

liver

gall bladder

duodenum

pancreas

stomach

intestine

rectum

bladder

b

Figure 2.16. Viscera of *Xenopus* and *Rana* compared, in a general dissection of each: ventral view. (a) *Xenopus*. (b) *Rana*

are short and occupy only the pectoral region alongside the heart and liver (Figure 2.16). The liver is also somewhat more extensive in *Xenopus* than in *Rana*. The most striking size difference, however, is in the hearts of these two animals. In *Xenopus*, the heart seems quite disproportionately large, and its ventricle protrudes posterior to the xiphisternal cartilage so that it is visible before the pectoral girdle is removed, whereas in *Rana* one does not see the heart until the girdle has been dissected away. The two hearts are compared in Figure 2.17. As will be mentioned later, (section 7), the roots of the aortic arches are also larger in *Xenopus* than in *Rana*.

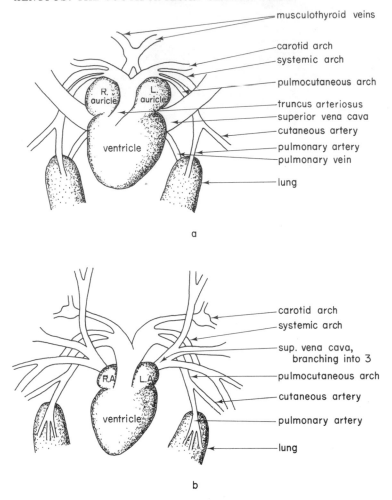

musculothyroid veins

carotid arch
systemic arch

pulmocutaneous arch

truncus arteriosus
superior vena cava
cutaneous artery

pulmonary artery
pulmonary vein

lung

R. auricle

L. auricle

ventricle

a

carotid arch
systemic arch

sup. vena cava,
branching into 3

pulmocutaneous arch

cutaneous artery

pulmonary artery

lung

R.A.

L.A.

ventricle

b

Figure 2.17. Heart and great vessels, compared in *Xenopus* and *Rana*: ventral views, from dissected specimens. (a) *Xenopus*. (b) *Rana*

The general topography of the gut is essentially similar in *Xenopus* and *Rana*, the only obvious difference being that the mesentery supporting the stomach and duodenum is more extensive in *Xenopus*. The urinogenital systems are also very similar in the two animals: there are only minor differences in the shapes of the kidneys and testes, which are slenderer in *Xenopus* than in *Rana*. The urinary bladder in *Xenopus* is smaller than that in *Rana*, and is spherical instead of bilobed. There are no seminal vesicles in *Xenopus*. These points are illustrated in Figures 2.18 and 2.19.

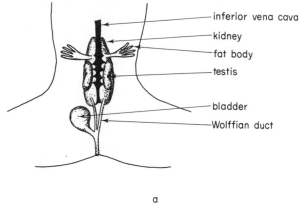

inferior vena cava
kidney
fat body
testis
bladder
Wolffian duct

a

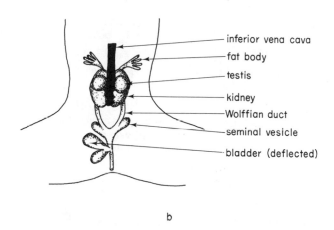

inferior vena cava
fat body
testis
kidney
Wolffian duct
seminal vesicle
bladder (deflected)

b

Figure 2.18. Male urinogenital system, ventral view.
(a) *Xenopus*. (b) *Rana*

7. The Vascular System

The following account is derived mainly from the detailed and beautifully illustrated descriptions by Millard (1941, 1945, 1949). She followed the complete sequence of development of both arteries and veins in *Xenopus*, from their earliest appearance in the larva before hatching right up to metamorphosis when the adult layout is acquired. The dissection of the finer branches of the blood vessels is difficult, even in the adult, and I have only been able to add some very elementary points from my own dissections and from comparisons with *Rana*.

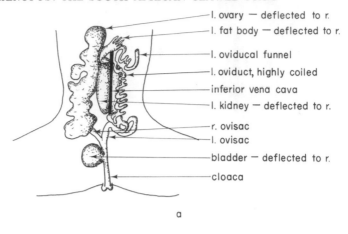

l. ovary — deflected to r.
l. fat body — deflected to r.

l. oviducal funnel
l. oviduct, highly coiled
inferior vena cava
l. kidney — deflected to r.
r. ovisac
l. ovisac
bladder — deflected to r.
cloaca

a

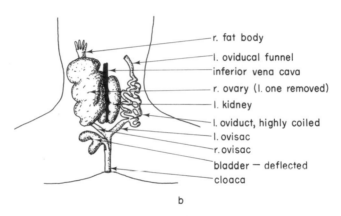

r. fat body
l. oviducal funnel
inferior vena cava
r. ovary (l. one removed)
l. kidney
l. oviduct, highly coiled
l. ovisac
r. ovisac
bladder — deflected
cloaca

b

Figure 2.19. Female urinogenital system, ventral view. (a) *Xenopus*.
(b) *Rana*

(a) The Venous System

The development of the veins was described by Millard in 1949. She noted that the omphalomesenteric, or vitelline, veins (equivalent to those similarly named in bird and mammal embryos) were the first to appear in the larva, at the 4 mm (pre-hatching) stage. These paired veins run along each side of the gut and enter the heart via the ducts of Cuvier (see Figure 2.20 (a)). Later they acquire anastomoses which loop round the gut, and one of these crosses dorsal to the developing liver (Figure 2.20 (b)). Subsequently the omphalomesenteric veins lose their connections with the heart, and part of the left one breaks up into capillaries in the liver, while its distal portion becomes the subintestinal vein (Figure 2.20 (c)). This vein is carried over to

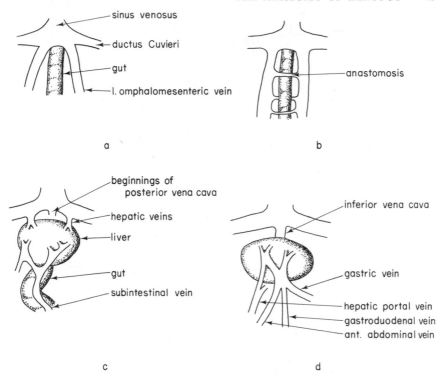

Figure 2.20 Stages in the development of the venous system of the gut in *Xenopus*. Redrawn after Millard (1949)

the right side with the duodenal loop, when the gut starts to coil, and it later acquires gastric and gastroduodenal branches from the stomach and duodenum. At metamorphosis, when the gut shortens, an originally small vein leading from the ileum to the right omphalomesenteric vein enlarges to become the hepatic portal vein (Figure 2.20 (d)). Thus the right omphalomesenteric vein persists, unlike the situation in other anurans, and it forms the proximal part of the hepatic portal system. Meanwhile the subintestinal vein (left omphalomesenteric) becomes the gastroduodenal vein (Figure 2.20 (d)). Millard points out that there is no subintestinal vein in other anurans whose vascular development has been described, although it is present in urodeles. She regards the development of the right omphalomesenteric vein in *Xenopus* as similar to that of the urodels *Salamandra*. In other anurans, this vein breaks up into capillaries in the liver, like that of the left side, and leaves no vestige in the adult.

A general comparison of the veins of the adult gut region in *Xenopus*, *Rana* and *Salamandra* is given in Figures 2.21 (a)–(c).

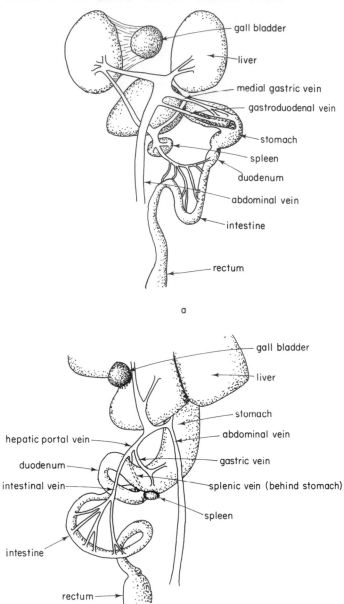

Figure 2.21. Veins draining the gut, in adult amphibians. Ventral views. (a) *Xenopus*, redrawn after Millard and Robinson, 1955. (b) *Rana*, redrawn after Whitehouse and Grove, 1947. (c) *Salamandra*, redrawn and simplified from Francis, 1934

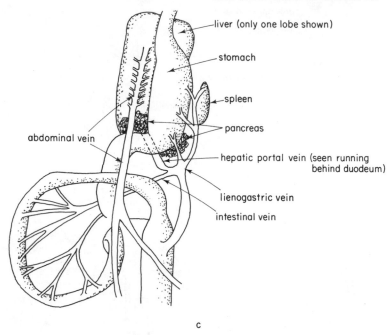

liver (only one lobe shown)

stomach

spleen

pancreas

abdominal vein

hepatic portal vein (seen running behind duodeum)

lienogastric vein

intestinal vein

c

Figure 2.21. (continued)

The anterior and posterior cardinal veins, serving the trunk region, are originally paired longitudinal vessels which according to Millard appear in the larva soon after hatching. They open on each side into an extensive sinus surrounding the pronephric kidney of the larva (Figure 2.22 (a)). The postcardinals soon begin to fuse in the midline, first at their caudal ends and then gradually cranial wards, thus forming the interrenal vein. At the same time a pair of lateral postcardinal veins gradually develops, extending from the pronephric sinus to the cloaca, where they join the interrenal vein in the midline (Figure 2.22 (b)). There is at first a network of small vessels linking the medial and lateral postcardinals on each side, at the level of the future mesonephric kidneys, and this network gradually merges to form a pair of large, continuous sinuses which surround the mesonephric ducts (Figure 2.22 (c)). At metamorphosis, however, this pair of sinuses is again modified into a network of fine vessels, and these now serve as the renal portal system for the mesonephric kidneys. (Figure 2.22 (d)). This portal system is very prominent in the adults I have dissected. The pronephric sinuses disappear, together with the pronephric kidney, at metamorphosis.

Throughout these changes in the trunk veins, part of the right postcardinal vein and the whole of the interrenal vein have remained joined to the midline

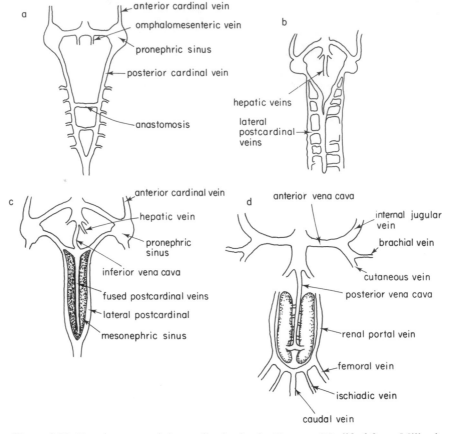

Figure 2.22. Development of the cardinal veins in *Xenopus*. Modified from Millard, 1949. (a) 7 mm tadpole. (b) 11 mm tadpole. (c) 24 mm tadpole. (d) Metamorphosis

trunk of the hepatic veins: these components become the posterior (inferior) vena cava (Figure 2.22 (d)). They also have ventral connections to the ventral (or anterior) abdominal vein (see below).

Millard regards the development of the veins of the trunk as essentially similar to that in other anurans, especially *Rana fusca*, described by Engels (1935). In both *Xenopus* and *Rana*, the ventral (anterior) abdominal vein develops from two lateral veins. The anterior parts of the veins fuse in the midline, while posteriorly, sections of one or other vein are lost, so that a continuous median vein eventually remains.

The development of the veins in the branchial region in *Xenopus* shows several specializations. Millard noted that, for instance, the external jugular veins (distal branches of the original anterior cardinal veins) are very prominent in the larva, and these drain the gill arches and the filter apparatus.

At metamorphosis, the jugular veins lose their branchial branches, but continue to receive blood from the hyoid and pharyngeal regions. There appears to be considerable variation in the course and nomenclature of branches of the external jugular veins in different Amphibia, but Millard sees some homologies between those of *Xenopus* and of urodeles (see Figure 2.23) rather than of other anurans. In *Xenopus*, some of the functions of the

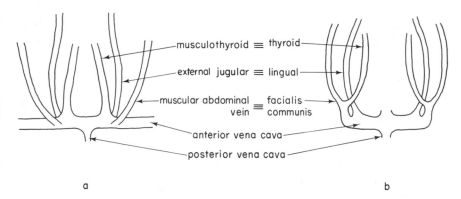

a   b

Figure 2.23. Homologies of the head veins in *Xenopus* and *Salamandra*, according to Millard, 1949. (a) *Xenopus*. (b) *Salamandra*

external jugular are taken over by the musculothyroid vein later, which first appears in young frogs and runs ventral to the larynx, draining the thyroid gland and the muscles of the hyoid apparatus. It is very prominent in the adult, where it appears immediately anterior to the roots of the aortic arches (Figure 2.17). It appeared larger on the right side than on the left, in specimens that I dissected, although Millard shows it larger on the left. Another unusual and important pair of veins in this region, which appear in *Xenopus* larvae, are the so-called 'muscular abdominal' veins. These drain the ventral surface of the head on each side and enter the ductus Cuvieri. The term 'abdominal' arises from the fact that these veins later acquire connections with muscles of the ventral body wall (rectus abdominis and sternohyoideus muscles) and may also have connections with the anterior abdominal vein.

The development of the veins in the head region was considered by Millard to show resemblances to urodeles. The anterior cardinal veins receive medial and lateral veins from the head and eye, and the medial vein later receives three branches from the brain: the anterior, medial and posterior cerebral veins. The anterior cerebral veins becomes the ophthalmic vein of the adult, and the medial cerebral vein becomes the orbitonasal vein. After metamorphosis the anterior vena cava on each side is formed from the shortened ductus Cuvieri, and it receives the internal jugular vein, which is

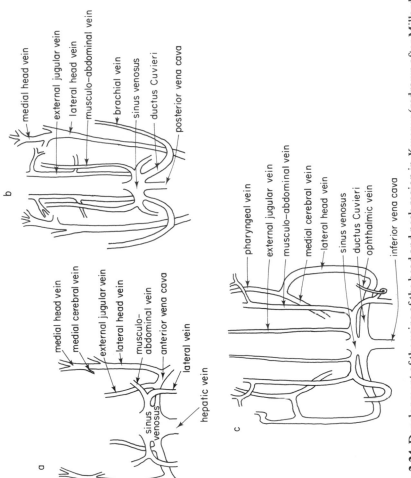

Figure 2.24. Development of the veins of the head and neck region in *Xenopus* (redrawn after Millard, 1949). (a) 8 mm tadpole. (b) 24 mm tadpole. (c) Metamorphosis

made up of the facial vein distally, the medial and lateral head veins and a portion of the anterior cardinal vein. These stages in development of the head veins are illustrated in Figure 2.24.

When one compares the layout of the adult venous trunks in *Xenopus* with those in *Rana*, the most striking difference is that there is only one main trunk of the anterior vena cava on each side in *Xenopus*, instead of the three that are present in *Rana*. In *Xenopus* all the main veins enter this single trunk (cf. Figure 2.25). There is also no musculothyroid vein in *Rana*, but

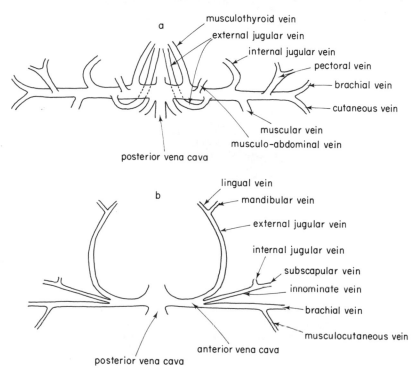

Figure 2.25. Main trunks of anterior vena cava. (a) *Xenopus*. (b) *Rana*

this animal has a common musculocutaneous vein instead of the separate muscular and cutaneous veins found in *Xenopus*. The cutaneous vein in *Xenopus* seems no less stout than its equivalent in *Rana*, despite the statements of earlier authors that the skin in *Xenopus* is unlikely to be an efficient respiratory organ owing to its inadequate blood supply compared with other anurans.

While regarding much of the venous system's development in *Xenopus* as typical of Anura, Millard listed a number of possibly neotenic features, resembling the Urodela. Among these are: the persistence of the right

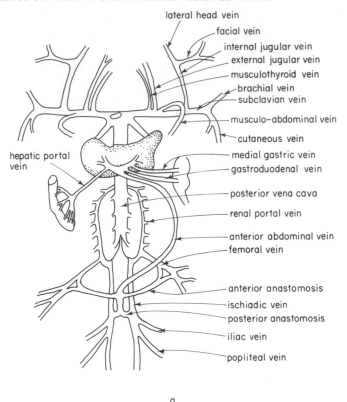

Figure 2.26. Adult venous system of *Xenopus*, compared with that of *Rana* and *Salamandra*. (a) *Xenopus*, redrawn from Millard, 1941. (b) *Rana*, redrawn from Borradaile, 1941. (c) *Salamandra*, modified from Francis, 1934. All in ventral view

omphalomesenteric vein and of the subintestinal vein, as mentioned earlier; the arrangement of the hind limb veins (see below); the layout of the external jugular veins and the origin of the head veins. The complete adult venous system in *Xenopus* is shown in Figure 2.26 (a), based on Millard's illustrations (Millard, 1941). It is seen here that the hind limb veins drain into the ischiadic vein, as in the urodele *Salamandra*, and not into the femoral vein as in *Rana* (Figure 2.26 (b)). There are also two anastomoses between the abdominal vein and the renal portal system, as in *Salamandra* (Figure 2.26 (c)).

(b) The Arterial System

Millard (1945) made an experimental as well as descriptive study of the development of the arteries in *Xenopus* larvae. Hers was apparently the

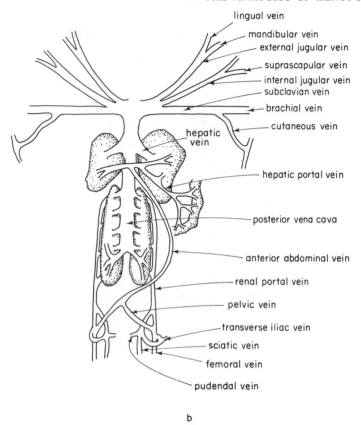

lingual vein

mandibular vein

external jugular vein

suprascapular vein

internal jugular vein

subclavian vein

brachial vein

cutaneous vein

hepatic vein

hepatic portal vein

posterior vena cava

anterior abdominal vein

renal portal vein

pelvic vein

transverse iliac vein

sciatic vein

femoral vein

pudendal vein

b

Figure 2.26. (continued)

first description of their development in any Aglossan amphibian, and she pointed out that *Xenopus* is of special interest owing to the fact that it does not develop internal gills specialized for respiration, as do other anuran larvae. *Xenopus* larvae have only external gills at first, and later a branchial filter apparatus internally which is thought to serve only for feeding purposes (cf. Chapter 5). As result, the branchial arteries have some unusual features.

According to Millard's observations, the heart, ventral aorta and first pair of aortic arches appear at the 4 mm pre-hatching stage, at the same time as the omphalomesenteric and anterior cardinal veins begin to develop (see above, p. 42). The first aortic arch is the hyomandibular arch. Very soon, before hatching, the third, or carotid arch appears too on each side. The fourth (systemic) arch is the next to appear, shortly after hatching, then rudiments of the second and fifth arches. The second or hyoid arch is only vestigial, but

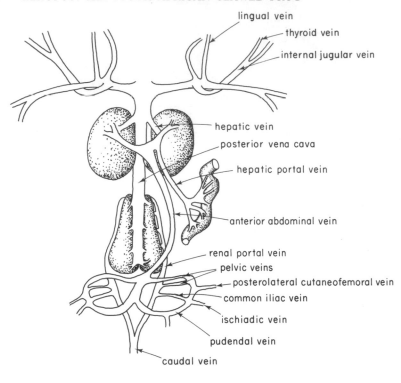

c

Figure 2.26. (continued)

the fifth persists and supplies the third gill arch (see Figures 2.27 (a)–(c)). Before the sixth (pulmocutaneous) aortic arch appears, the first and second arches begin to degenerate. Only the distal end of the first arch remains, as the palatine artery of the adult, and no vestige at all remains of the second aortic arch. At the stage when the mouth has broken through (7 mm larva), the sixth aortic arch is seen developing round the fourth gill arch and sending a branch to the lung. Later, in 8 to 10 mm larvae, the external gills shrink and the branches from the aortic arches supplying them also dwindle. Finally, three complete aortic arches, the third, fourth and fifth are left, while part of the sixth remains as the pulmonary artery and it also sends branches to the filter apparatus. Surprisingly, the third (carotid) arch and not the fourth (systemic) as in other tetrapod vertebrates, enlarges at this stage and becomes the major route for blood from the ventral to the dorsal aorta. This may be because in *Xenopus* the fourth arch becomes specialized for

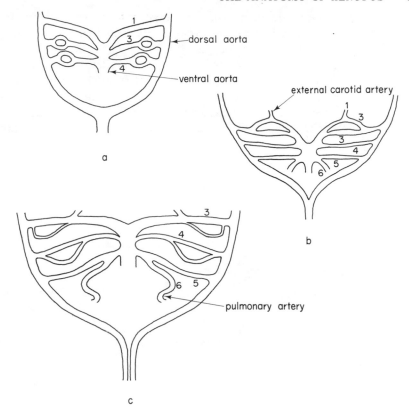

Figure 2.27. Diagram illustrating the development of the aortic arches in *Xenopus*. Redrawn from Millard, 1945. (a) 5 mm larva, (b) 8 mm larva, (c) 10 mm larva

supplying the filter apparatus: it breaks up into a number of capillaries in the branchial region and also receives branches from the sixth arch. Thus an elaborate vascular network supplies the filter apparatus (Figure 2.28), and in view of this, Millard felt that the apparatus must have at least a transient function in respiration. This has been denied by other authors, though, because blood from this region in *Xenopus* is returned directly to the anterior cardinal veins, and not to the dorsal aortae, as there is no efferent branchial arterial system as in fish and in other larval amphibians. Millard points out, however, that respiration by other means than the branchial system is unlikely to be adequate, since there is only a very small cutaneous branch off the sixth aortic arch in the larva, not adequate to serve for cutaneous respiration, and the lungs at this stage are only simple sacs with relatively little internal surface area.

Millard was able to throw a little light on the functions of individual aortic arches in *Xenopus* larvae by experiments in which she destroyed certain arches by cautery at early stages. She found that larvae were unable to survive destruction of one or both third aortic arches, confirming that these are of major importance in carrying blood from the ventral to the dorsal aorta. Loss of the fourth arch on one side did allow survival, however: this was partly because a ductus caroticus persisted on that side to carry blood via the third arch to the subclavian artery (see Figure 2.29). Loss of the

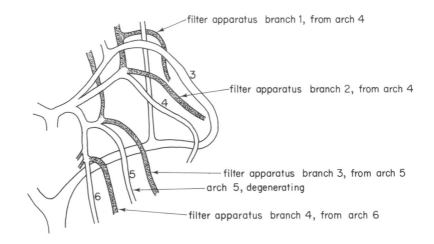

Figure 2.28. Arterial supply to filter apparatus. Modified from Millard, 1945. Left-hand side only drawn: diagrammatic. Shaded vessels are branches to filter apparatus

fourth arch on both sides was fatal, however, probably because of its essential supply to the filter apparatus: this supports the idea that the filter apparatus is respiratory as well as being concerned in feeding. Destruction of either or both sixth arches was also fatal. This indicates that a blood supply to the lungs may be essential for respiration, but it could also support Millard's emphasis of the possible respiratory function of the filter apparatus, since branches of the sixth arch also supply this apparatus, as we have seen. The cutaneous artery is also no doubt essential for life, whether or not the skin has any important respiratory function at this stage.

In the adult arterial system, it is difficult to trace all the fine branches in the head and neck region, but Millard (1941) produced excellent illustrations of the arterial supply to all parts of the body and Millard and Robinson (1955) show the main features. Figure 2.30 compares the layout of the main arterial trunks in *Xenopus* and *Rana*. These trunks are noticeably more robust in *Xenopus*, their size being proportionate to that of the much larger heart than in *Rana*. Several other points of difference between the arteries of *Xenopus* and *Rana* were noted by Millard, and I have been able to confirm most of these by dissection. First, the laryngeal artery is a branch of the external carotid (very difficult to find, since it enters muscle close to its origin). Millard describes it as larger in the male than in the female, but it is no thicker, as far as I have seen. Secondly, there is a separate, stout ventral muscularis branch of the external cartoid, which is not found either in the

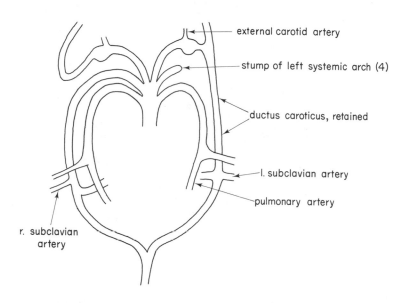

external carotid artery

stump of left systemic arch (4)

ductus caroticus, retained

l. subclavian artery

pulmonary artery

r. subclavian artery

Figure 2.29. Persistence of ductus caroticus on left-hand side after cautery experiment in which the 4th aortic arch has been destroyed. Redrawn from Millard, 1945

closely rated genus *Pipa*, according to Millard, or in *Rana* which has a slender musculocutaneous artery in this position. A further feature described in *Xenopus* by Millard is the small size of the cutaneous branch of the sixth aortic arch. However, in specimens that I have dissected, this artery is quite as stout as that of *Rana*. The cutaneous artery has no auricular branch, however, as it has in *Rana*: this is presumably because of the absence of any external ear-drum in *Xenopus*. A further special feature of the sixth aortic arch in *Xenopus* is that its pulmonary branch has only a single stem with several small branches coming off it, instead of the three main stems seen in *Rana*. Several of these points are compared in Figures 2.30 (a) and (b).

In the arteries of the trunk, *Xenopus* also shows some differences from other anurans. The coeliaco-mesenteric artery has a lower origin than in

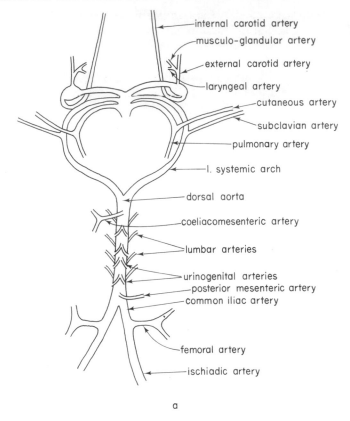

internal carotid artery

musculo-glandular artery

external carotid artery

laryngeal artery

cutaneous artery

subclavian artery

pulmonary artery

l. systemic arch

dorsal aorta

coeliacomesenteric artery

lumbar arteries

urinogenital arteries

posterior mesenteric artery

common iliac artery

femoral artery

ischiadic artery

a

Figure 2.30. Adult arterial system: main vessels. (a) *Xenopus.*
(b) *Rana*

Pipa or *Rana*: it arises from the single abdominal aorta instead of from the left thoracic aorta (Figure 2.30). In the pelvic region, branches of the femoral artery supply the seminal vesicle of the male or the uterus of the female, and these regions do not receive branches from the iliac artery as in other Anura. This may be correlated with the specializations of the venous system in this region which were noted earlier (cf. Figure 2.26 and p. 50).

The only other consistent difference noted by Millard in the arterial system of *Xenopus* as compared with *Rana*, is that the occipito-vertebral artery arises from the subclavian and not directly from the dorsal aorta. Millard found more individual variations in the arterial pattern in *Xenopus* than have been described for other anurans: but this may simply reflect the

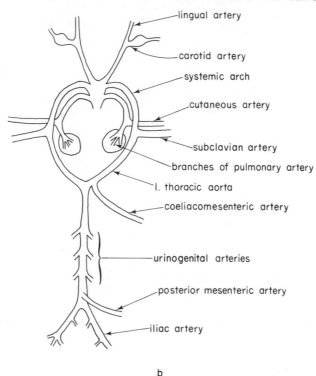

lingual artery

carotid artery

systemic arch

cutaneous artery

subclavian artery

branches of pulmonary artery

l. thoracic aorta

coeliacomesenteric artery

urinogenital arteries

posterior mesenteric artery

iliac artery

b

Figure 2.30. (continued)

greater detail of her observations and the large number of animals on which they were based.

This survey of the anatomy of *Xenopus* shows up some of the physiological specializations in this animal, besides its differences from other anurans and intriguing resemblances to urodeles. We shall go on in the next chapter to consider experimental work on the physiology of this animal.

3

Physiological Studies on *Xenopus*

Much of the physiological work on *Xenopus* in the 1930's and 1940's was concerned with its reproduction. The ready response of this animal to injections of gonad-stimulating hormones was found to be of practical use to clinicians in the diagnosis of pregnancy: so for a time its reproductive physiology became a predominant interest. We shall consider the use of *Xenopus* in pregnancy testing and the work on its reproduction in the latter part of this chapter. But first, some mention must be made of the wider variety of physiological work that has been carried out on *Xenopus* since it was first exploited as an experimental animal in the scientific laboratories of South Africa.

1. Some General Features of the Physiology

The earliest investigations carried out on *Xenopus* in physiological laboratories reflect the special interests of the observers, rather than any peculiar features of *Xenopus* as compared with other amphibians. For instance, Zwarenstein and his collaborators, whose work we shall deal with first, had for some time been concerned with the general problem of how energy reserves are maintained in cold-blooded vertebrates. *Xenopus* was readily available and easy to keep in the laboratory, so they carried out some of their work on this animal.

(a) Factors affecting Blood Sugar Levels in Xenopus

Bosman and Zwarenstein (1930) showed that *Xenopus* normally maintains a fairly constant level of blood sugar: 44 mg %. But this is affected by changes in temperature: they found that the maximal blood sugar level reached after injections of glucose was higher at higher temperatures. Zwarenstein and Bosman (1933) showed that the level of glucose tolerance in the blood also depended on pituitary activity: it was higher if the pituitary gland had previously been removed (by the operation known as 'hypophysectomy'). This indicated that the pituitary was perhaps involved as a stimulator of the pancreas in controlling blood sugar levels. As one would expect, injections of pancreatic insulin caused a drop in blood glucose and phosphate levels, but this drop was not so marked in hypophysectomized animals as in normal ones (Shapiro & Zwarenstein, 1934: Schrire, 1939). The blood phosphate is in equilibrium with energy-rich reserves of phosphate in muscle, in the form of creatine: so it is interesting that Shapiro and Zwarenstein noted a gradual decrease in muscle creatine in hypophysectomized animals. More recently, responses to insulin have been compared in adult and larval *Xenopus* by Hanke and Neumann (1972). They find that the response is greater in adults, but that glycogen storage is promoted by insulin in both adults and larvae.

Environmental factors which affect the output of certain pituitary hormones may also cause changes in blood glucose levels. For instance, if *Xenopus* is placed on a light background, its skin melanophores contract under hormonal control and it becomes paler: at the same time the blood sugar level drops to about 33 mg % (Bosman & Zwarenstein, 1934). When these animals are transferred to a dark background, the melanophores expand in response to an outflow of melanophore-stimulating hormone (MSH) from the pars intermedia of the pituitary, and at the same time the blood sugar rises. Conditions of stress, too, affect blood sugar and phosphate levels: the stress hormone, adrenalin, produced by the adrenal glands in response to stimulation by adreno-corticotrophic hormone (ACTH) from the pituitary, causes a drop in blood phosphate levels at first (Schrire, 1939), but this is followed by a slow rise to higher levels than normal. More recent work by Hanke and Leist (1971) confirms that in larvae, too, there is a rise in blood sugar and also in the levels of glycogen in the liver and other tissues, after ACTH injections during the metamorphosis period. Conditions of stress also cause an increase in respiratory rate, so the rise in blood phosphate may be a part of this syndrome (Hogben, 1934). Presumably the heart rate also speeds up: but there appears to be little data on normal heart-rates in *Xenopus*. Jolly (1934) compared the rate of beating of the lymph hearts with that of the heart: he recorded lymph heart rates of 56 beats/min at 16·5 °C and 75 beats/min at 20·5 °C. These were about 2·6 times as rapid as the main heart-beat at these temperatures. The beat of the lymph hearts varied with circumstances, however: they underwent phases of inactivity, as well as phases of very rapid beating when the animal made vigorous muscular movements. Jolly thought that the lymph hearts might serve some role in assisting respiration, but this interesting idea has not been either confirmed or refuted since. There is still considerable controversy about the mechanisms of respiration in *Xenopus* (see Emilio & Shelton, 1974). Lymph hearts are not a peculiarity of *Xenopus*: in all anurans there are at least two pairs, one situated below the scapulae and opening into the subscapular veins, and the other pair at the end of the urostyle opening into the femoral veins.

(b) Colour Adaptations in Response to Background

One of the reasons why the colouring of different species of *Xenopus* is difficult to define is that it varies in response to changes of background. It has been known for some time that the response is due mainly to the action of the melanophore-stimulating hormone, MSH, produced by the pars intermedia of the pituitary gland. Some of the important ground-work on the nature and effects of MSH was carried out on *Xenopus* in the 1930's. Slome and Hogben (1929, 1934) first showed that colour changes in response to background were under hormonal and not nervous control, by demonstrating that the changes still occurred when the nerve supply to an area of skin had been destroyed. Hypophysectomy, however, blocked these colour changes. Hogben and Gordon (1930) succeeded in obtaining an extract from the pituitary gland of *Xenopus* that caused expansion of the melanophores in the skin. This was the first isolation of MSH from *Xenopus*. In the next decade, it became a standard method for the quantitative assay of MSH, to observe the degree of expansion of melanophores in the foot web of *Xenopus*, where they could easily be seen *in vivo* under a dissecting microscope (Landgrebe & Waring, 1944).

There are other agents which have effects similar to MSH. Adrenalin, for example, causes dispersal of pigment in the melanophores, so that the animals

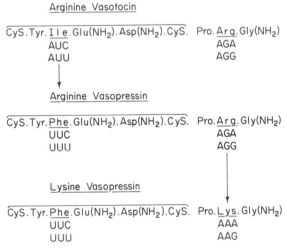

Figure 3.1. Comparison of the structures of pressor hormones in vertebrates—after Heller and Pickering, 1970. The nucleic acid base triplets coding for those amino-acids that differ in the three hormones are shown. It can be seen that a change of a single base, caused by a mutation, could convert vasotocin into vasopressin, and arginine vasopressin into lysine vasopressin

darken in colour (Burgers, 1956). It is now thought that catecholamines such as adrenalin always play an essential role in the darkening reaction, since the reaction is abolished by agents which inhibit the uptake and metabolism of catecholamines (Brouwer, 1970, 1972). A secretion from the skin glands, called serotonin, also causes dispersal of pigment (Veerdonk, 1960). The opposite effect, contraction of the melanophores, occurs in response to a hormone known as melatonin. This is thought to be secreted by the pineal gland in *Xenopus* (Bagnara, 1963), since removal of the pineal abolishes the 'paling' response to a light background, and administration of melatonin can restore the response (Charlton, 1966). But the highest concentrations of melatonin have been found, not in the pineal but in the eye tissue (Baker & Hoff, 1971). Other hormones which may antagonize the effects of MSH are the 'pressors' from the posterior pituitary tissue (Hogben & Gordon, 1930). For example, vasopressin, whose other effect is to increase blood pressure by constricting capillary vessels, also causes contraction of melanophores.

It is interesting to note, in passing, that *Xenopus* like other amphibians has a posterior pituitary hormone which appears to be an evolutionary precursor of vasopressin: namely arginine-vasotocin (Follett & Heller, 1964). The composition of this hormone and its relationship to mammalian vasopressin have been set out by Heller and Pickering (1970)—see Figure 3.1. Acher and coworkers (1964) showed that the vasotocin of *Xenopus* is chemically and pharmacologically similar to that of *Rana* and of other Amphibia.

A number of workers have studied the conditions governing the release of MSH from the pituitary gland in *Xenopus*. It is evidently partly under the control of the hypothalamus of the brain, for it has been seen that nerves supply the pars intermedia from the paraventricular area of the hypothalamus, and transection of this nerve tract abolishes the darkening response. Burgers, Imai and van Oordt (1963) showed that changes in MSH content of the pituitary occurred gradually over a period of 48 hours after animals had been changed from one background to another. Pituitaries of light-adapted animals contained over four times as much MSH as those of dark-adapted animals whose melanophores had expanded. This suggests that much MSH had been released into the blood stream of the dark-adapted animals. The fact that these animals nevertheless still had some MSH in their pituitaries, might mean that there was a rapid synthesis of hormone after dark-adaptation. Cohen (1967) carried the investigations further by studies on the ultra-structure of the MSH-secreting cells. He saw secretory vesicles and endoplasmic reticulum (indicative of protein synthesis) in the cells of light-adapted animals, but not in those of dark-adapted animals. So his evidence goes against the idea that there is any immediate synthesis of MSH during dark-adaptation. The renewal of the hormone stores must therefore either be very gradual, or must take place at times when it is not being released into the blood stream. Thornton (1971b) has confirmed by measurements of the MSH content of the pituitary, that there is a rapid fall in concentration during dark-adaptation: he also found that there was a rapid resynthesis of MSH in animals adapting to a light background.

(c) Visual Responses

That the eyes also play a part in the responses of *Xenopus* to background is suggested by their high content of melatonin (see above) and was confirmed by Vilter (1946). He rotated the eyes 180° and found that although these experimental animals showed the same response as controls to a light background, their response to a dark background was much slower and less intense. Vilter explained this result by the fact that overhead light was now striking the part of the retina that had originally been dorsal and normally perceived changes in the background (see Figure 3.2). So the effect of the overhead light tended to inhibit responses to a black background, since the animals took some time to adjust to the disorientation of their retina. On the other hand, overhead light would tend to reinforce the change to paler colour in response to a light background.

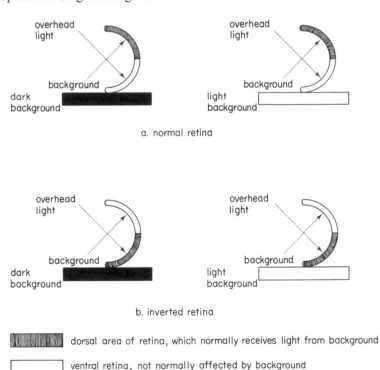

Figure 3.2. Diagrams to illustrate the effects of Vilter's eye rotation experiments on the background response

Further visual functions have been studied in the eye of *Xenopus*: for example, colour vision. Burgers (1952) tested the ability of adult frogs to distinguish different combinations of coloured light, and found that blue and green were not distinguished from each other, though these could both be distinguished from colours at the other end of the spectrum. Blue, green and yellow were each confused with one particular shade of grey, which suggests that none of these was perceived as a colour quality. Red also did not seem to be perceived as a colour. A more precise study was carried out by Cronly-Dillon and Muntz (1965) on the wavelengths of light perceived by *Xenopus* tadpoles. They showed that these animals were sensitive to light of wavelengths well into the red end of the spectrum. Their maximum sensitivity varied according to the background: on a bright background the sensitivity was maximal at 630 mμ, but on a dark background there were two maxima of sensitivity, at 460 mμ and 560 mμ.

The eye is also concerned in equilibration in *Xenopus*. Zoond and Rimer (1934) found that if both the membranous labyrinth of the inner ear, and the eye, were removed, *Xenopus* frogs were unable to respond to rotation on a turntable. But removal of the labyrinth only, leaving the eyes intact, allowed a response to the turntable to occur, even *in the dark*: the mechanism of this reaction is not known.

(d) Pharmacology of Other Organs

Functions of some other organs in *Xenopus* have been studied by isolating them *in vitro*. To give a few examples: the physiology of nerve conduction was studied on isolated single fibres from the sciatic nerve of *Xenopus*, by Schoepfle, Atkins and Schafer (1965). They found that the rate of change of electrical potential in the nerve membrane was decreased—hence the rate of conduction of a nerve impulse was reduced—in the presence of iodoacetate. This is a feature of nerve conduction in other vertebrates also.

Ion transport across the skin of *Xenopus* has also been studied, by Callaghan (1953). He found that in isolated sacs of leg skin, the rate of transport of ions increased with temperature, up to a maximum at 25 °C. Temperatures above 25° caused autolysis and death of the skin. This is also the maximum temperature that the intact animals will tolerate. The permeability of the skin to water has also interested physiologists, since amphibians which emerge on land at metamorphosis show a lowered body water content. Despite remaining aquatic, *Xenopus*, like other Anura, shows decreased permeability of the skin to water after metamorphosis (Schultheiss, Hanke & Metz, 1972). This change in permeability is less marked in hypophysectomized animals, but is restored to normal by administration of prolactin. Bentley (1969) has noted that vasotocin does not increase the rate of transport of water across the skin of either *Xenopus* or urodeles, *in vitro*, but it does increase the transport of sodium ions. This may be an equally vital function in fresh-water animals.

Another organ of *Xenopus* that has been used in isolation for pharmacological experiments is the stomach. Zwarenstein (1953) was able to show that it secretes pepsin, although the presence of this enzyme in other amphibians has been questioned.

After this fairly broad sample of the kinds of physiological work that have been carried out on *Xenopus*, we must now go on to consider the major line of interest in its reproduction, and how this has been applied to pregnancy testing in humans.

2. The Use of *Xenopus* for Pregnancy Tests

There are rival claimants to the discovery of the *Xenopus* test for pregnancy. Hogben, Charles and Slome (1931) were the first to show that administration of extracts of the anterior pituitary gland to female frogs causes them to lay eggs. Later it was shown that the active factors in the pituitary were the gonadotrophins (gonad stimulating hormones) which cause maturation and shedding of eggs in the female, or sperm in the male. The essential hormone for initiating ovulation (shedding of eggs from the ovary) is 'LH', or luteinising hormone. This is so named because in pregnant female mammals it causes the ruptured follicles which remain after ovulation to transform into corpora lutea: these in their turn secrete hormones essential for the maintenance of pregnancy. Appreciable concentrations of LH are found circulating in the blood during early pregnancy, and are excreted in the urine. Hence it is possible to test for pregnancy in women, by injecting hormone extracted from the urine into female frogs and seeing if these respond by laying eggs. The response can be quantitated to a limited extent, but not precisely, because there is much individual variation in the reactions of animals to the hormone. *Xenopus* has an advantage over other species, in responding at any time of year and not only during the breeding season. Now, however, much more sensitive and quantitative methods are available for detecting pregnancy in its early stages and for assaying LH.

In 1932, Shapiro and Zwarenstein reported their discovery of the *Xenopus* test for pregnancy to the Royal Society of South Africa (see Zwarenstein & Duncan, 1944). At the same time the test had been discovered by Bellerby in London, but he did not publish his findings till 1934. These pioneer reports led to a spate of work aimed at improving the test procedure: this work has been reviewed by Weisman and Coates (1944a). There were several difficulties in extracting the hormone in purified form from urine: at first alcohol was used as precipitant, but this caused some fatalities among the frogs injected. Later, an absorbent kaolin was used to concentrate the hormone (see Zwarenstein & Duncan, 1944). Eventually a chromatographic extraction method was devised by Milton (1946). It was also found that the laboratory conditions under which the frogs were kept greatly influenced their sensitivity to the hormone (see section 3).

Besides normal pregnancy, certain clinical conditions such as malignant tumours of the placenta (e.g. chorionepitheliomata and hydatidiform moles) and testicular tumours associated with gynaecomastia, could also be detected by the *Xenopus* test, since in these conditions there are also high concentrations of LH in the blood and urine. All these tests, however, depend for their accuracy on the sensitivity and reliability of the response by the *Xenopus* frogs, which as we have seen is somewhat variable. So the *Xenopus* test fell out of use as soon as more sensitive and accurate assay methods for gonadotrophins were devised. It was not long before Fried and Rakoff (1955) that hyperaemia of the ovary of the rat occurred in response to much lower doses of LH than those required to produce ovulation in *Xenopus*. Since then, immunological methods for detecting the hormone have been developed to a very high sensitivity. The haemagglutinin-inhibition test (Politzer 1963; Politzer & Simpson 1963) is used nowadays and gives 98 % correct diagnoses, as against 87 % with the *Xenopus* test. Hence, and most fortunately for developmental biologists, there is no longer a great demand for *Xenopus* by pregnancy diagnosis centres, and the South African populations of these useful frogs are not in so much danger of becoming depleted as they were in the 1940's. At the same time, developmental biologists have to thank the clinicians for discovering many useful points about the successful husbandry and breeding of *Xenopus* in the laboratory. We will now go on to consider some of these points in more detail.

3. Reproduction in *Xenopus*

While the reproduction of *Xenopus laevis* has probably been studied more extensively in the laboratory than that of any other amphibian species, its breeding habits in the wild have not been so well documented recently. We have to look back to some of the early descriptions of this species, dating from before it became a very popular laboratory animal, for accurate information about wild populations of *Xenopus* in South Africa.

(a) Breeding of Xenopus in the Wild

Berk (1938) gives the breeding season of *Xenopus laevis* in the Cape region of South Africa as extending from July to September. This agrees with Hey's statement (1949) that breeding starts in late winter and extends to early spring, though Hey puts the starting point at August rather than July. Even in this relatively short season, each female normally lays two or three batches of eggs. At more northern latitudes, despite the hotter climate, *Xenopus* breeds later. In the Transvaal, for instance, Kalk (1960) reports that breeding extends from September to January, starting at the onset of the rainy season. This suggests that moisture, rather than temperature, is the governing environmental factor. Winter and spring in the Transvaal are relatively dry, and the nights are cold, whereas the summer, though hot, is the more humid season. For a fully aquatic animal, the ability to find plenty of water in the ponds is the most important determinant of its breeding season.

According to Waring, Landgrebe and Neill (1941), *Xenopus* females do not shed by any means all of their eggs from the ovaries in a breeding season. The ovaries regress gradually after breeding, however. In the Cape the regression period is from September onwards (Berk, 1938) and the growth period is from February till July. Gitlin (1939, 1942) found that the oviducts also showed a seasonal variation in weight corresponding to that of the ovaries, and that the variations in both these organs were governed by pituitary hormones.

In the male of *Xenopus*, the most obvious seasonal variation is that of the secondary sexual characters. A mature male in the breeding season shows clearly the black, sticky hairs on the inner surfaces of the digits, palms and forearms which help to grip the female during coupling (cf. Figure 1.3 on p. 6). The presence of these hairs is an indication that sperm are being produced in the testis. Injections of LH to induce the shedding of sperm are rarely successful with males which do not already show blackened digits.

(b) Breeding Induced in the Laboratory

Under laboratory conditions, the seasonal pattern of gonadal growth and breeding activity in *Xenopus* is usually disturbed, though Weisman and Coates (1944a) claimed that they could keep females in good breeding condition for several months. Alexander and Bellerby (1935) were the first to notice a deleterious 'captivity effect' causing the ovaries of the females to regress unless very careful attention was paid to feeding and other conditions. Landgrebe (1939) stressed the need to avoid overcrowding of the animals. This is an even more important precaution now that town water supplies are often heavily chlorinated. Possibly one of the reasons why the addition of green algae to the water has been found to promote breeding in *Xenopus* (Bles, 1905; Hey, 1949; Savage, 1965) is that this helps to remove chlorine. In our own laboratory, some inexplicable deaths of *Xenopus* after no symptoms except lethargy, have been attributed to the high chlorine content of the water. It is clearly important to have the water well aerated, as well as providing the animals with abundant food.

For success in producing fertile eggs in the laboratory, the usual procedure is to give both sexes an adequate dose of luteinizing hormone while their gonads are in good condition. The animals must then be left in the dark for several hours, as undisturbed as possible, to induce them to couple and to lay. Weisman and Coates (1944b) tested this point by giving identical doses of hormone to two groups of *Xenopus*, one of these then being kept in bright light and the other group in the dark. They found that very few of those in the light responded to the hormone, and that their response was delayed compared with the ones in the dark.

(c) The effects of Hormones on Gonads and Sexual Behaviour

(i) The Female.   The normal cycle of growth and regression of the ovary in *Xenopus* is governed, as in all other vertebrates, by oestrogenic hormones. These are secreted by the ovarian tissue under the influence of pituitary gonadotrophins. In *Xenopus*, two oestrogenic hormones, oestradiol and oestrone, have both been identified in extracts from the ovaries (Gallien & Foulgoc, 1960). The hormones are present in the ovaries from an early age, one month after metamorphosis. When ovaries were compared at stages before ovulation, just after injection of gonadotrophins, or immediately after egg-laying, the titres of the two hormones varied: there was a twofold increase in concentration of oestradiol, 14 hours after gonadotrophin injection. In mature females, both the hormones were also found to be circulating in the blood. Recently Redshaw and Nicholls (1971) have shown that the main site of synthesis of these hormones is the follicle cells.

Both ovarian growth and the maintenance of ovarian hormone output are dependent on stimulation by the pituitary gland, as we have already noted. Thus, hypophysectomy causes regression of the ovaries (Bellerby & Hogben, 1938) and injection of pituitary extracts, even from quite different sources such as the sheep (Gitlin, 1942), promote ovarian growth and maturation of the ova in *Xenopus*. Although LH is normally the stimulus to ovulation, Shapiro (1939) showed that a wide variety of steroids would induce *Xenopus* to ovulate. He found ten active compounds, among both the androsterone series (related to male hormones) and the pregnane series (Figure 3.3). Most of these had a delta-4-3-keto constitution. These steroids were also effective on hypophysectomized females. They were not all effective on other amphibian species, however: possibly because these were less able than *Xenopus* to respond to hormones outside their breeding season. Later workers have found that while luteinizing hormone is normally adequate to produce ovulation in freshly caught females of *Xenopus*, those which have been in captivity for some time may need first to be brought into breeding condition by injections of follicle-stimulating hormone (FSH). Parker, Robbins and Loveridge (1947) had success with a preparation from sheep anterior pituitary glands, and recently Dodd (pers. comm.) has used a commercial serum gonadotrophin with success, for both males and females that had been in captivity for some months. These FSH preparations also enhance the secondary sexual characters: thus they induce enlargement and reddening of the cloacal labia in the female, and darkening of the digits and forearms in the male (Berk & Shapiro, 1939; Berk, 1939; Gallien, 1948).

There has been some controversy as to which regions of the anterior

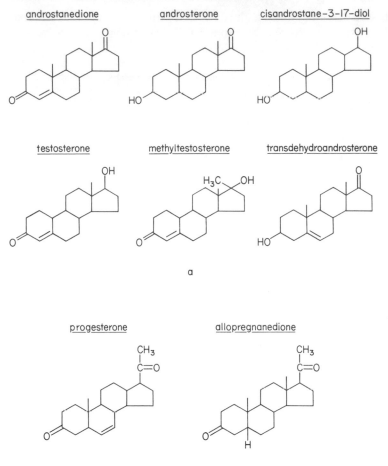

Figure 3.3. Structures of hormones which cause ovulation in *Xenopus*, according to Shapiro (1939). (a) Androsterone series. (b) Pregnane series

pituitary lobe in *Xenopus* secrete the two gonadotrophins, FSH and LH. Studies by Evennett and Thornton (1971) of extracts from both halves of the lobe suggest that LH activity is widely distributed, and that the type 2 basophil cells which mature a few months after metamorphosis (Kerr, 1965, 1966) secrete both FSH and LH. The distribution of these cells is shown in Figure 3.4.

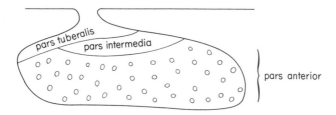

Figure 3.4. Diagram to show distribution of Type 2 basophil cells in the pituitary gland (sagittal section). Redrawn from Kerr, 1965. Circles represent distribution of the cells

(ii) The Male. The male *Xenopus* shows a rapid response to injections of gonadotrophin, both secondary sexual characters and spermiogenesis being affected. Robbins, Parker and Bianco (1947) found that gonadotrophins either from the anterior pituitary of the sheep or from the human chorion caused the appearance of spermatozoa in the cloaca in only $1\frac{1}{2}$–2 hours after injections into the dorsal lymph sac of the frog. The males reacted to much lower doses of hormone than were needed to induce ovulation in females, but Hobson (1952) found that the sperm are difficult to detect because they are sometimes non-motile. Gallien (1948) had found the same difficulty. In further work Hobson and Barr (1965) observed that *Xenopus* responds equally to the two components of gonadotrophins: the FSH and the interstitial cell stimulating factor (ICSH). This is unlike *Rana* which is relatively insensitive to FSH. Testicular hormones seem ineffective in inducing breeding in laboratory *Xenopus*: both testosterone and testosterone propionate have been tried without success (Berk, 1939).

The clasping and mating behaviour of *Xenopus* males can be elicited dramatically by doses of gonadotrophic hormones. This fact has been utilized in investigating the ability of animals without either vision or smell to find their mate and to couple successfully. Surprisingly, it has been found that neither of these faculties is essential. After hormone injections, eyeless *Xenopus* males may take longer to find their mate and may at first adopt a slightly abnormal coupling position, but this is soon corrected and normal coupling takes place (Berk, Cheetham & Shapiro, 1936). Removal of the nasal organs from either or both sexes is also without effect on the coupling behaviour of the male in response to gonadotrophin injections (Shapiro,

1937). This response does, however, require the presence of gonads: Shapiro (*loc. cit.*) found that whereas hypophysectomized animals showed normal behaviour after injections of anterior pituitary extracts, those whose gonads were removed and their pituitaries left intact showed no mating response. Shapiro concluded that in *Xenopus* (as in many other vertebrates), mating behaviour depends on an interaction between the pituitary and the gonads, in which the gonads play a positive role.

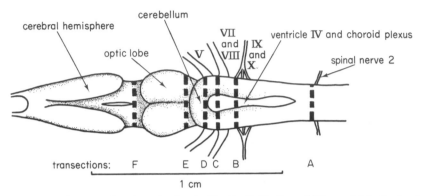

Figure 3.5. Diagram showing the levels (A, B, C, D, E and F) at which the brain was transected, to test the control of the clasp reflex. (From Hutchison and Poynton, 1968)

The tendency of the breeding male *Xenopus* to clasp the female with its forearms (or to clasp the experimenter's finger if he inserts it between them!) appears to be an entirely reflex response. Hutchison and Poynton (1963) tried the effects on the clasp reflex of transecting the brain of male and female *Xenopus* at various levels. It was found that transections at the midbrain level reduced the intensity of the clasp reflex in the male, and transections between the hindbrain and the spinal cord abolished the reflex altogether. If, however, the transection was made at the level of the anterior medulla (B, Fig. 3.5), there was a stronger clasp reflex in response to hormone injections than in normal, unoperated males. More surprisingly, the reflex was also elicitable in females which had been subjected to this level of transection, although unoperated females never show any such reflex. Evidently the severance of the anterior medullary connections with the midbrain had the effect of masculinizing the female's behaviour. In mammals, it is known that certain cyclic sexual activities in females are initiated in the hypothalamus region of the forebrain. It is possible that in *Xenopus* females this region is responsible for inhibiting activities that are characteristic of males. If so, however, one would expect the appearance of a clasp reflex in females after transections of the midbrain, as well as after anterior medullary transections, since either operation must destroy connections between the brain stem and the forebrain floor (Figure 2.9 above). We need much more information on the central control of sexual behaviour in amphibians before critical experiments can be designed to elucidate these results.

4. Effects of Hormones on Sexual Differentiation

It is well known that the development of sexual organs and the differentiation of sex in vertebrates is dependent not only on sex chromosomes but also on hormonal control. Some years ago Witschi (1938) showed that when amphibian larvae are grafted together in pairs (an operation known as 'parabiosis': Figure 3.6), hormonal factors from the male member of a pair cause partial masculinization of the other individual if this was initially a female. But the reverse effect, feminization of a male parabiont by its female partner, is not observed under these conditions. So it has been concluded that the hormones of genetically female amphibians are neutral as regards control of sexual development, but that there are active factors in genetic males which are able to masculinize immature gonadal tissue, enhancing the differentiation of its medullary (future testicular) portion and suppressing the development of its cortical (ovarian) portion. This theory does not rule out the possibility that genetically female tissue may produce local factors enhancing cortical development and suppressing medullary development in its own gonads, but any such factors are evidently not capable of influencing another animal in parabiosis, so they are not carried in the blood stream and may not, therefore, be regarded strictly as 'hormones', which are by definition blood-borne agents.

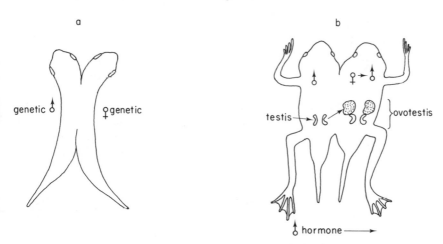

Figure 3.6. Parabiosis of amphibian larvae, and its effect on sexual differentiation. (a) Larvae at time of operation. (b) Metamorphosed frogs. Female is masculinized by a hormone from the male

Chang and Witschi have carried out extensive experiments on sex reversal in *Xenopus* larvae. Chang (1953) found that, besides parabiosis operations, testis grafts were also capable of causing masculinization of genetically female larvae. The ovarian tissue of the host's gonad was suppressed by a testis graft, and there was hypertrophy of the medullary part of the gonad. Chang and Witschi (1955) went on to test the effects of adding male hormones (androgens) to the medium in which the larvae were reared: but this procedure did not cause any masculinization of female gonads. The androgens did, however, cause male secondary sexual characters to develop in otherwise female animals: after metamorphosis they showed blackening of the digits and forearms, and attempted to couple with other, normal females! It is difficult to explain this lack of effect of androgens on the gonads, in view of their other positive effects. Later, however, Chang and Witschi (1956) reported that androgen treatment was capable of restoring to normal the development of a testis which had been partially feminized by oestrogen treatment (see below). So although androgens do not seem able to initiate masculinization of the gonads, they evidently do have some influence on their differentiation later.

Another paradoxical aspect of the sex-reversals obtainable in *Xenopus* was observed both by Chang and Witschi (1956) and by Gallien (1953). They found that although in parabiosis experiments there is never any feminization of a male partner by its female parabiont, treatment of larvae with oestrogenic hormones added to the water causes 100 % conversion of males into females. Whereas the normal sex ratio is 1:1 males:females, when larvae are reared from hatching to metamorphosis in 0·6 mg/ml of oestradiol benzoate (Gallien, 1953) all of them turn out to be females. When these are crossed with normal males, 50 % of the females give rise to all-male offspring: this is because their chromosome constitution is that of the male, i.e. ZZ sex chromosomes, instead of WZ as in normal females. (Table 3.1 makes this point clear). This test was in fact one of the ways in which it was proved that in *Xenopus*, as in urodeles, the male is the 'homogametic' sex, producing gametes all of one kind, with a Z chromosome. Similar breeding experiments were carried out later with females which had been masculinized by testis grafts (Mikamo & Witschi, 1964), and these showed clearly that the genetic female is heterogametic, giving rise to two kinds of gametes, with either W or Z chromosomes. The offspring from crosses of the masculinized females with normal females are found to consist of 75 % females (25 % WW and 50 % WZ) and 25 % males (ZZ). Table 3.2 illustrates this test. Clearly the presence of either one, or two W chromosomes produces a female, but it needs two Z chromosomes to produce a genetic male.

Mikamo and Witschi (*loc. cit.*) claimed that they were able to masculinize one genetically WW larva, by means of a testis graft. This claim needs to be substantiated by more cases, with cytological observations on the chromosome constitution of the animals, before one may conclude whether or not all the effects of the two Z chromosomes can be mimicked by hormones from the testis. There is the dilemma, anyway, that a testis must first be produced, before it can be used as a graft and exert its masculinizing influence. So far, we know of no other way of producing a testis initially, except by having two Z chromosomes in the cells of the larval gonad.

Table 3.1. Sex ratios in offspring of *Xenopus*

Normal situation:

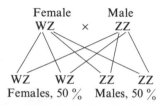

Experimental situation:

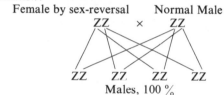

It can be seen that the work on *Xenopus* has thrown light on many important physiological features of lower vertebrates. It has contributed to our understanding of the roles of hormones in controlling blood sugar levels and colour changes, and the relative roles of hormones and visual functions in certain aspects of behaviour. The research into its reproductive physiology that stemmed from work on pregnancy testing has resulted in a

Table 3.2. Sex ratio in offspring of masculinized females

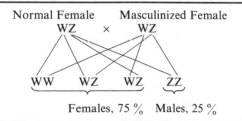

number of interesting observations on the control of sexual behaviour and sexual differentiation in this animal.

We must now go on to consider, in a separate chapter, the course of events by which male and female gametes are produced, and the process of fertilization.

4

Gametogenesis, Fertilization and the Initiation of Embryonic Development

Having considered the hormonal and chromosomal factors that govern development of the gonads and of sexual characters in *Xenopus*, we must now go on to deal with the development of the germ cells: ova in the female and spermatozoa in the male. Both the formation of ova (oogenesis) and the formation of spermatozoa (spermatogenesis) from the primitive germ cells involve meiotic, or reduction divisions so that the chromosome number is halved. The sequence of steps in both processes is set out in Figure 4.1: it is the same as in all other vertebrates.

In the following account, we shall consider spermatogenesis briefly first and will then go on to discuss the many interesting features that have come to light from cytological and biochemical studies of developing oocytes in *Xenopus*.

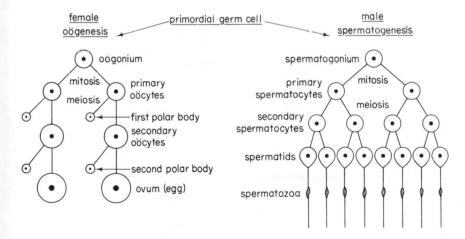

Figure 4.1. Diagram of the stages of gametogenesis

1. Spermatogenesis

The spermatozoa of *Xenopus* are difficult to study when mature, since as we noted earlier, a proportion of those that are emitted are non-motile and therefore not easily found. A probable reason for this is that *Xenopus*, unlike other amphibians, has no seminal vesicles for storage of sperm after they have left the testis. As a result, the sperm accumulate in the bladder, where some may succumb to toxic materials in the urine.

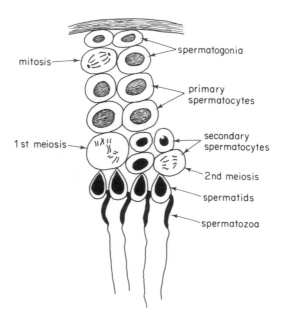

Figure 4.2. The sequence of cell types seen in a cross-section of a seminiferous tubule. Redrawn after Balinsky, 1970

The early stages of spermatogenesis in the testis of *Xenopus* have, however, been followed without difficulty. They were first described in detail by Firket (1963), in animals which had been injected with chorionic gonadotrophin so that the time course of events could be followed accurately. Firket found that within 24 hours after an injection of hormone, all mature sperm had disappeared from the seminiferous tubules of the testis and a new cycle of spermatogenesis had begun. It appears from Firket's account that this process in *Xenopus* is essentially the same as in other amphibians. Figure 4.2. shows diagrammatically the cell types that would be recognizable in a histological section of a seminiferous tubule in which spermatogenesis was taking place. Firket noted that just as in mammals, two types of spermatogonia can be distinguished: pale cells which give rise to spermatocytes, and darker cells which remain at the periphery as stem cells.

Ultrastructural studies of spermatogenesis in *Xenopus* have been made by Kalt (1972), and by Reed and Stanley (1972) who emphasized several points of interest. First, they noted that the primordial germ cells of both sexes have large, lobated nuclei (cf. Figure 4.3 (a)). These had already been remarked on in oogonia by Al-Mukhtar and Webb (1971). Similar lobated germ cell nuclei have also been seen in *Rana*. That the spermatogonia in *Xenopus* are indeed derived from the primordial germ cells (see section 3) is supported by Reed and Stanley's observations of dense, granular material associated with the mitochondria, in both germ cells and spermatogonia. They believe that

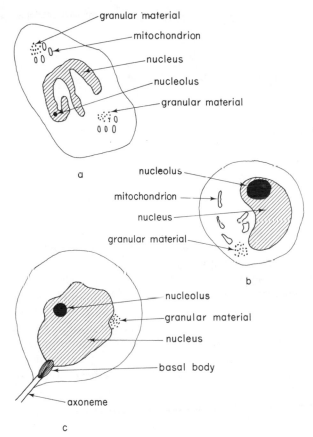

Figure 4.3. Fate of granular material during spermatogenesis in *Xenopus*, redrawn from photographs of Reed and Stanley, 1972. (a) Primordial germ cell. (b) Spermatogonium. (c) Spermatid beginning to form

this material later gives rise to the large chromatoid body of the sperm head (cf. Figures 4.3, 4.5, 4.6 and 4.7). A special feature of the primary spermatocytes in *Xenopus* are the synaptonemal complexes in the nucleus, thought to be evidence of chromosomal pairing and crossing over (Figure 4.4). The spermatocytes also form a vesicular membrane structure in their cytoplasm, resembling a Golgi complex. This is cast off, together with other cytoplasm, at the spermatid stage.

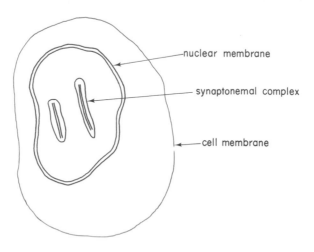

Figure 4.4. Primary spermatocyte, showing synaptonemal complexes. Redrawn from Reed and Stanley, 1972

Some of Reed and Stanley's electron micrographs are reproduced in Figures 4.5–4.9, to show the sequence of structural changes. It is particularly interesting that the spermatids develop within groups of so-called 'follicle' cells: nutritive cells that are thought to be comparable to the Sertoli cells of the mammalian testis. The elongation of the spermatids may be helped by the expansion of microtubules in the cytoplasm which are orientated along their long axis. The developing flagellum has a ciliary rootlet-like structure at its base (cf. Figure 4.6 (d)). Reed and Stanley also confirm, by phase-contrast microscopy, that the sperm head in *Xenopus* is spiral in shape, unlike other amphibian sperm so far examined. There have been no comparisons of the motility of *Xenopus* spermatozoa with those of other amphibians, but possibly the spiral head enables them to swim faster. In *Xenopus* the pelvic amplexus position of the male (cf. Figure 1.1 above) means that the sperm have further to travel in order to reach the eggs than in other anurans whose pectoral amplexus position brings the male and female cloacae close together. It is said that the passage of the sperm towards the eggs in *Xenopus* is facilitated by mucus on the skin, but this has not been tested experimentally.

Figures 4.5–4.9
Ultrastructure of spermatogenesis in *Xenopus*. From Reed and Stanley, 1972

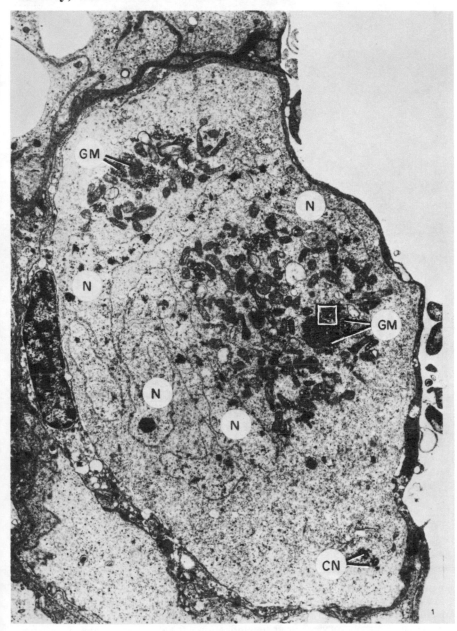

Figure 4.5. Primordial germ cell in the testis. Linear magnification × 6,000. Note the granular material (GM), centrioles (CN) and lobulated nucleus (N)

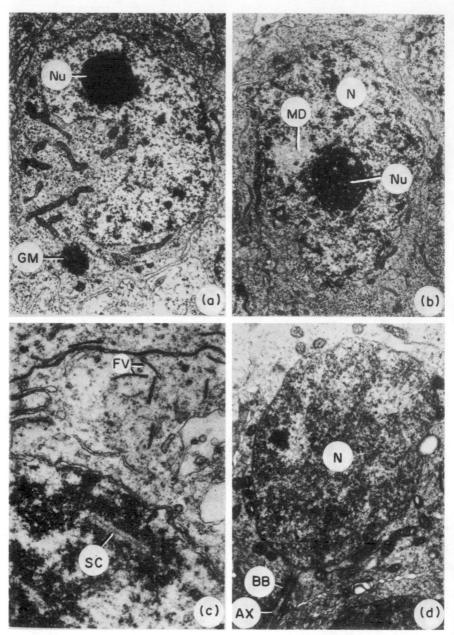

Figure 4.6. (a) Early spermatogonium, with large nucleolus (Nu) and granular material (GM). Linear magnification × 6,000. (b) Mature spermatogonium. Nucleolus now contains a mass of thread-like structures. MD = zone of medium density. Linear magnification × 6,000. (c) Synaptonemal complexes (SC) and flattened vesicles (FV) in a primary spermatocyte. Linear magnification × 6,000. (d) First signs of spermiogenesis. Basal body (BB) and axonemal thread (AX) are seen at base of nucleus (N) in a young spermatid

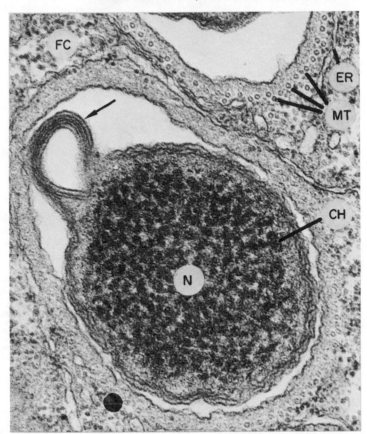

Figure 4.7. Nucleus of a spermatid. Linear magnification × 60,000. Adjacent to it is a follicle cell (FC) with microtubules (MT) which run parallel to the long axis of the spermatid. Excess spermatid membrane, possibly surrounding the beginnings of the acrosome, is seen at arrow. CH = chromatoid body.

2. Oogenesis

The developing oocyte, because of its large size and its cytoplasmic stores of yolk and other reserves, has received far more study than the spermatocyte. There have been a number of ultrastructural studies of oogenesis in *Xenopus*. One of the earliest of these was by Balinsky and Devis (1963) who followed the course of yolk- and pigment-formation in some detail. They observed that oocytes of up to 150 μ diameter (stage 'A', cf. Grant, 1953) still have no yolk or pigment, and that yolk platelets first appear in oocytes of 300 μ diameter (stage 'B'), at the periphery of the cell. In oocytes of 325–450 μ diameter (stage 'C'), there is a gradual progression in the appearance of yolk, from

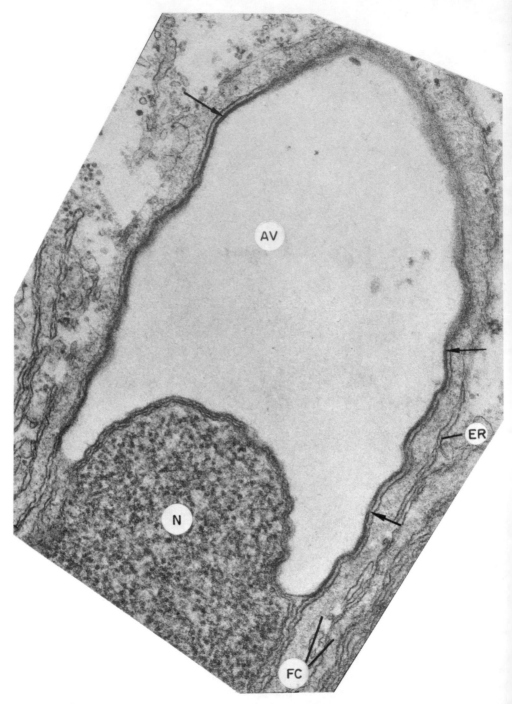

Figure 4.8. Large acrosomal vesicle (AV) overlying the spermatid nucleus as it develops. Note follicle cell (FC) with endoplasmic reticulum (ER). Linear magnification × 45,000

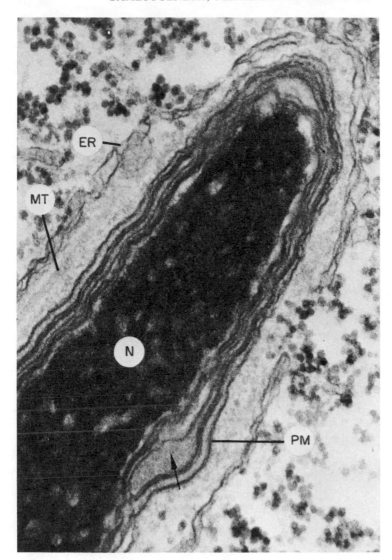

Figure 4.9. Mature spermatid in which the acrosome is flattened over the surface of the nucleus. It is seen best at arrow, where the plasma membrane of the follicle cell also shows (PM). Linear magnification × 90,000. MT = microtubules

the periphery to the centre. Pigment granules begin to appear at stage 'D', 475–750 μ diameter, when the oocytes are filled with yolk. The mature oocytes (stage 'E') are from 850–1000 μ diameter and have besides pigment a layer of cortical granules at the surface. These granules, which seem to originate

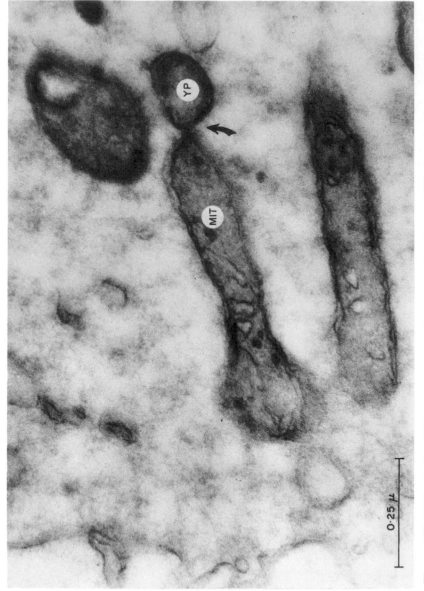

Figure 4.10. Yolk platelet (YP) shown on the end of a mitochondrion (MIT), in a developing oocyte. From Balinsky and Devis, 1963

from the Golgi body, are destined to break down at fertilization. Figure 4.11 shows some of the stages in oocyte development described by Balinsky and Devis.

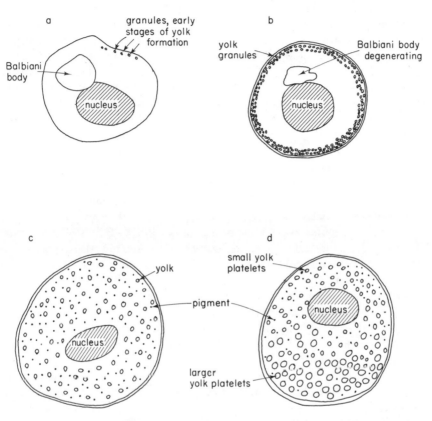

Figure 4.11. Stages in oogenesis, redrawn after Balinsky and Devis, 1963. (a) Stage A, (b) Stage C, (c) Stage D, (d) Stage E

Considerable controversy has centred round the origin of the yolk platelets in amphibian oocytes. There is a region near the nucleus, called the Balbiani body after its first discoverer, which appears to be the centre for yolk synthesis: yet the first fully formed platelets are seen at the periphery of the cell, as we noted above. Evidently as they form, the platelets must move outwards, reaching maturity when they get to the periphery. The Balbiani body is an accumulation of mitochondria, endoplasmic reticulum (organized ribosomes) and free ribosomes. It may receive material from the nucleus too, since Al-Mukhtar and Webb (1971) observed granules in the pores of the nuclear membrane, which were thought to be passing out to the mitochondria (cf. also Reed and Stanley's description of granular material in spermatogonia, mentioned on p. 77). It is in association with the mitochondria that yolk platelets seem to originate. Balinsky and Devis (1963) observed spherical bodies containing concentric membranous layers, situated on the ends of the mitochondria and apparently being formed by mitochondrial activity (cf. Figure 4.10 above). These membranous bodies have since been identified as precursors of yolk platelets. Other observers had shown similar pictures of yolk platelets forming on the ends of mitochondria (Ward, 1962; Wartenberg 1964). Balinsky and Devis followed in some detail the deposition of yolk in the membranous vesicles as they gradually transformed into yolk platelets. Since the structural details can only be seen with the high resolution of the electron microscope, it is not surprising that in earlier observations based only on light microscopy, it was sometimes thought that yolk platelets actually turned into mitochondria: or, alternatively, that mitochondria were transformed into yolk platelets. Electron microscopy has shown that the two are separate entities throughout the development of the oocyte.

Some of the vesicles which are found in the Balbiani region in oocytes of *Xenopus* give rise first to granular bodies, according to Balinsky and Devis. These are of two kinds: fine-granular bodies which later give rise to peripheral yolk, and bodies with coarser granules which later form pigment. Wartenberg (1964) also observed that pigment formed in granular bodies situated at the ends of the mitochondria in *Xenopus*, and Hope, Humphries and Bourne (1964a, b) described yolk forming from vesicular bodies, which they thought were derived from mitochondria, in other amphibian species. So evidently both pigment and yolk in *Xenopus* and in other amphibians are synthesized in close association with mitochondria.

There have been some interesting recent investigations on the origin of the proteins from which the yolk in the oocyte is formed. Wallace and Dumont (1968) estimated that over 80 % of the protein nitrogen and 90 % of the protein phosphorus of the amphibian egg was contained in the yolk: so it is obviously a very important store of protein. The sources of this protein have been traced by radioactive and immunological labelling methods, in *Xenopus* females which were injected with gonadotrophins to stimulate the growth of the oocytes. After an injection of gonadotrophin, a lipophosphoprotein which Wallace and his colleagues named 'serum lipophosphoprotein', or 'SLPP', is synthesized in the liver of the female toad. It passes into the blood serum and thus to the ovary, where it is taken up, apparently undegraded, into the protein of the developing oocytes. The passage of this SLPP has been traced either by labelling it with radioactive phosphorus (^{32}P) or by preparing a fluorescent antiserum which can be used to label it in sections of tissues. A surprising discovery has been that implants of the ovarian hormone β-oestradiol will cause synthesis of SLPP in the liver, in males as well as females. The response occurs also in hypophysectomized animals: in fact even liver slices isolated *in vitro* will respond to ovarian hormone by synthesizing SLPP. The uptake of the SLPP into oocytes has also been followed *in vitro* (Wallace & Jared, 1969). It is taken up in preference to other serum proteins, by a micropinocytosis mechanism, which enables large molecules to enter the cells intact (Dehn & Wallace, 1973). Finally, the SLPP can be detected in the yolk platelets when these form.

Not all yolk platelets in the amphibian oocyte are of the same size. In mature oocytes of *Xenopus*, Czołowska (1969) described three concentric zones: a peripheral zone with small yolk platelets, a subperipheral zone with compact, intermediate-sized platelets and an internal zone with loosely-arranged, large platelets. Also in this internal zone, islands of strongly basophilic cytoplasm are seen: this is thought to be the 'germ plasm' which passes into the germ cells during embryonic stages. As we shall see below, it appears that in at least some amphibians a continuous line of 'germ plasm' is inherited, via the female, as each new generation of oocytes forms. This phenomenon has aroused new interest in Weisman's classical 'germ plasm theory', which fell out of fashion when the inheritance of characters through nuclear genes was discovered. It now seems that his theory may yet hold, in the sense that some heritable factors are segregated into the germ cells and do not enter the rest of the body.

This remarkable germ plasm in *Xenopus* deserves further comment.

3. Germ Plasm and Germ Cells in *Xenopus*

Bounoure (1934) first drew attention to the presence in early amphibian embryos of a basophilic cytoplasm which accumulated in the endoderm and appeared to pass into the germ cells before these migrated to the gonads. Blackler (1958) studied this 'germ plasm' in embryos of *Rana, Bufo* and *Xenopus*. He found that it had staining properties indicating that it was rich in RNA. In blastulae of *Xenopus*, cells containing this cytoplasm were found in the floor of the blastocoel, very near the ventral surface (cf. Figure 4.12). These cells could still be seen in the ventral endoderm at gastrula and neurula stages. By stage 40, when the larva was hatched and had begun to feed, however, the cells containing germ plasm were found to have migrated dorsally. A little later (stages 44–47), similar cells, but no longer with recognizable germ plasm, were found in the genital ridges, where the gonads were forming. It was not possible in *Xenopus* to trace completely the fate of the germ plasm by staining methods, but Blackler (1960, 1962) proved that the original ventral cells which contain the germ plasm do indeed colonize the gonads, by experiments in which he exchanged grafts of these cells between embryos of two subspecies, *Xenopus l. laevis* and *X. l. victorianus*. The *X. l. laevis* strain used was a mutant with only one nucleolus instead of the

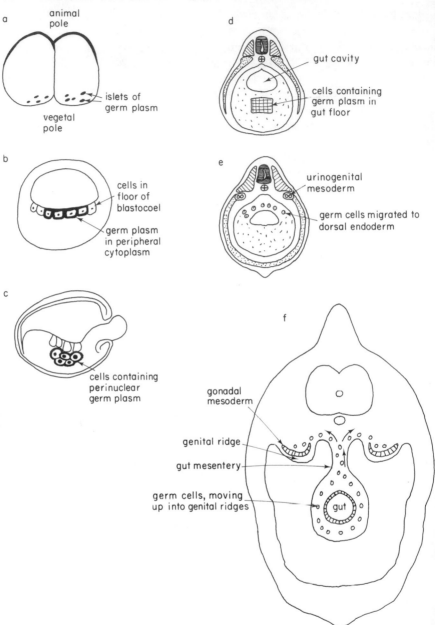

Figure 4.12. Diagrams to show the position of the germ plasm and germ cells at stages of development in *Xenopus*. (a) 2-4 cell stage. (b) Blastula, transverse section. (c) Gastrula, longitudinal section. (d) Neurula stage, transverse section. (e) Early larva (stage 40), transverse section. (f) Larva, stage 47, transverse section

normal two, so that the cells of host and graft in each case could be distinguished by the number of their nucleoli. It was found that most of the germ cells which appeared later in the gonads were of graft origin. In confirmation of this, matings of female hosts with males of their own subspecies, produced eggs which were nevertheless of the size and colouring typical for the graft's subspecies, not the host's. This technique of grafting germ cells from a mutant to a normal frog, or from one subspecies to another, has many possible applications in genetic studies on amphibians.

In more recent work, the germ plasm in *Xenopus* has been recognized at very early cleavage stages, at the ventral ends of the blastomeres (Figure 4.12 (a)). It seems to originate from the basophil cytoplasm in the centre of the mature oocyte which was described by Czołowska (see above). Buehr and Blackler (1970) showed that removal of some of the basophil cytoplasm at the 2-cell or 4-cell stage, by pricking the ends of the blastomeres, led to partial sterility of the animals that developed from these operated embryos. The number of mature germ cells found in the gonads was reduced, in proportion to the amount of basophil cytoplasm lost. Thus it seems clear that this basophil cytoplasm is essential both for the differentiation and for the migration of the germ cells. Recently Ikenishi *et al.* (1974) and Tanabe and Kotani (1974) have inactivated the germ plasm in *Xenopus* by ultra-violet irradiation: they describe several ultrastructural charges in it resulting from irradiation.

It is intriguing to realize that, since gametogenesis is already beginning in the larval stages, there is only a very short period during which the germ plasm cannot be seen in *Xenopus*: i.e. from the time that the germ cells approach the gonads (when this cytoplasm loses its staining properties, as we saw above) until the next generation of oocytes reaches maturity, when the plasm reappears. This period is only about $\frac{1}{10}$ th of the animal's life cycle, and in *Xenopus* lasts during larval stages 44 to 55, i.e. about 30 days. At all other times in females, the RNA-rich cytoplasm can be recognized, and it is fair to assume that it passes into the oocytes of each new generation. Wallace, Morray and Langridge (1971) suggest that this plasm may carry supernumerary gene copies derived from the maternal nucleus. As we shall see in the next section, there is much replication of DNA in the developing oocyte.

4. Nucleic Acid Synthesis in the Oocyte

(a) Sources of DNA and RNA

The developing oocytes of Amphibia accumulate both DNA and RNA in their cytoplasm, in addition to having both these forms of nucleic acid in the nucleus. The source of cytoplasmic DNA and RNA must be either the oocyte's own nucleus, which is large but whose contents seem to be disperse, or breakdown products from the nuclei of other cells such as the follicle cells. Neither DNA nor RNA has been traced with any certainty either from the follicle cells or from the oocyte's nucleus into the cytoplasm, however. In some other animals such as the insects, we do know that DNA is passed into the oocyte from follicle or 'nurse' cells, but this has yet to be proved for Amphibia. Pores 400 Å in diameter were observed in the nuclear membrane by Callan and Tomlin (1950), and it is now thought that small molecules of DNA or RNA could pass through to the cytoplasm. Possibly some poly-nucleotides pass out and become the precursors of cytoplasmic DNA and RNA: or possibly the passage takes place mainly during nuclear divisions, when the nuclear membrane breaks down.

(b) Synthesis of RNA in the Oocyte Nucleus

There have been a number of studies on the synthesis of RNA in the nucleus of the oocyte of *Xenopus*. It has been known for some time that

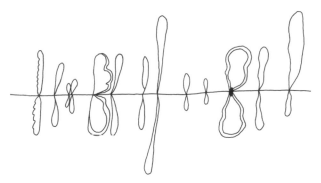

Figure 4.13. A lampbrush chromosome of an amphibian oocyte: diagrammatic. Redrawn after Waddington (1962)

nuclei of amphibian oocytes during the prophase of the first meiosis have 'lampbrush' chromosomes (Callan, 1952) with many side loops (see Figure 4.13). These loops have been shown by autoradiographic labelling methods to be active sites of synthesis of RNA (Gall, 1963), and the base composition of the RNA indicates that it could be a transcribed copy of the chromosomal DNA. Synthesis of RNA in the lampbrush loops is far more active than in nuclei of somatic cells: Davidson, Allfrey and Mirsky (1964) estimated that the RNA/DNA ratio in lampbrush chromosomes is 100 times that in the chromosomes of somatic cells. It has also been estimated that 90 % of the RNA synthesized at the loops is ribosomal RNA (rRNA). This rRNA must pass into the cytoplasm, if it is to contribute to the formation of ribosomes. Scheer (1973) has shown that there is a steady increase of rRNA in the oocyte cytoplasm, and has calculated that at 'mid-lampbrush' stage, 2.62 rRNA molecules pass through per 'nuclear pore complex', in which the nucleolus is also involved. As we shall see later, a reserve of ribosomes is built up in the oocyte cytoplasm which is sufficient to carry the embryo through its development up to the gastrula stage, when new ribosomes begin to be formed after the reappearance of nucleoli.

(c) The Nucleoli

Another peculiar feature of the nucleus in the developing amphibian oocyte, besides its 'lampbrush' chromosomes, is its large number of nucleoli. Coggins (1973) has observed that oogonia already have more nucleoli than somatic cells, and later in oocytes, some 1500 nucleoli are formed per nucleus, according to Perkowska, MacGregor and Birnstiel (1968). These extra nucleoli are destined to disappear at the second meiotic division, however, when the oocyte becomes an egg, leaving the early embryo with no nucleoli at all until the late blastula stage. From then on, each diploid somatic cell has just two nucleoli, which are formed at constrictions called 'nucleolar organizers' on one pair of chromosomes. It has been argued that the loops on the lampbrush chromosomes of oocytes, where rDNA and rRNA seem to originate, are comparable to nucleolar organizers, but this is still a matter for controversy (see MacGregor, 1965). The DNA cores of the oocyte nucleoli later seem to arise from an extrachromosomal source (see section (d) below).

An important point that has now been demonstrated with certainty, by work on anucleolate mutants in *Xenopus*, is that the nucleoli control the synthesis of ribosomal RNA (rRNA). So the large number of nucleoli in the oocyte are presumed to function mainly to ensure that an adequate reserve of ribosomes is formed, which can carry out all necessary protein syntheses in the early embryo, during the period after these oocyte nucleoli have disappeared and before the somatic cells acquire new nucleoli. In anucleolate mutants of *Xenopus*, no nucleoli appear at all in the somatic cells, and hence no new ribosomes are formed. The embryos do not survive beyond the tailbud stage, presumably because their supply of ribosomes derived from the oocyte is insufficient to carry out the protein syntheses required for development beyond this stage. These anucleolate embryos also lack any nucleolar organizer region on their chromosomes (Khan, 1962).

There is considerable variation in the size of the nucleoli, even in normal diploid cells of *Xenopus*. Occasionally there may be only one, double-sized nucleolus, probably formed by fusion of the original two (Esper & Barr, 1964). In oocytes, Van Gansen and Schram (1972) found even more variations in form of the nucleoli. They distinguished three main types: first, large, definitive nucleoli; secondly a smaller type which they called 'micronucleoli' and thirdly an intermediate type, the 'nucleolar bodies', which looked like amalgamations of several micronucleoli, possibly a stage in the formation of a definitive nucleolus. All of these types were shown by histochemical methods to be synthesizing rRNA. The ultrastructure of oocyte nucleoli has been reviewed by Kezer, MacGregor and Schabtach (1971). Some of Van Gansen and Schram's pictures are redrawn in Figure 4.14.

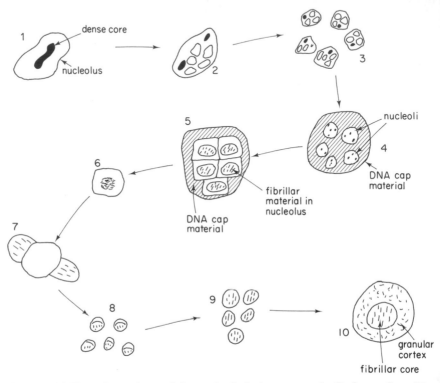

Figure 4.14. Transformations of the nucleoli during oogenesis. Redrawn from Van Gansen and Schram, 1972. Stages are: 1. oogonium; 2, 3, leptotene oocyte; 4. pachytene oocyte; 5. late pachytene oocyte; 6, 7, diplotene oocyte; 8, 9, diplotene previtellogenic oocyte; 10. vitellogenic oocyte

Hay and Gurdon (1967) showed that homozygous anucleolate mutants of *Xenopus* do possess micronucleoli, in both oocytes and somatic cells, but they have no definitive nucleoli or nucleolar bodies. Animals heterozygous for the mutation have only half the normal number of definitive nucleoli in the oocyte, and only one instead of two nucleoli in somatic cells. They also have only one nucleolar organizer region (Barr, 1966). Wallace and Birnstiel (1966) showed by molecular hybridization tests of their DNA against normal rRNA, that only half the normal complement of rDNA sequences was present in the heterozygotes. Such hybridization tests against rRNA, which can also be done *in situ* on the chromosomes and viewed in electron micrographs, were used to confirm that the DNA responsible for rRNA synthesis (i.e. the rDNA) is located in the nucleoli when they are present. This had already been deduced, from the lack of rRNA synthesis and hence of new ribosomes, both in anucleolate mutants and in normal embryos prior to the reappearance of nucleoli before gastrulation. Miller and Knowland (1970) have described yet another mutant, the 'partial nucleolate'. In this mutant there is one small-sized nucleolus in each somatic cell, and half the number of rDNA sequences as in the heterozygous anucleolates: i.e. a quarter the number in normal cells.

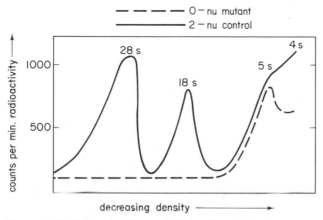

Figure 4.15. Diagram of a typical profile obtained by centrifugation of a purified RNA extract on a sucrose density gradient. Samples from normal and anucleolate mutant *Xenopus* embryos are compared. Redrawn after Brown and Gurdon, 1966

Analysis of the molecular sizes of RNA synthesized in these nucleolar mutants shows deficiencies in the synthesis of the two main ribosomal categories, 28s and 18s RNA (Figure 4.15). These are reduced proportionately to the reduction in rDNA. Some 5s rRNA is synthesized, however, and also DNA-like RNA (dRNA). Gurdon and Ford (1970) have shown that the dRNA can become attached to pre-existing ribosomes in the mutants, and that these can then clump together as polysomes and carry out some peptide synthesis. At the tailbud stage in homozygous anucleolates this dRNA becomes degraded, however. It is interesting that although in all these mutants the synthesis of ribosomal proteins is reduced or absent, paralleling their deficiencies in rRNA, their total synthesis of other proteins identifiable by electrophoretic analysis is no lower than in normal embryos up to the tailbud stage (Hallberg & Brown, 1969: see Figure 4.16).

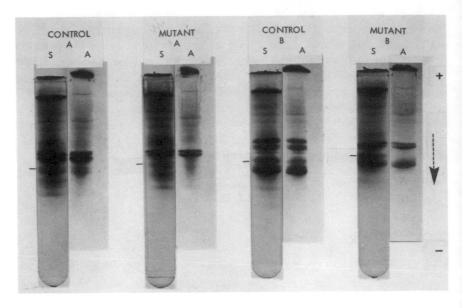

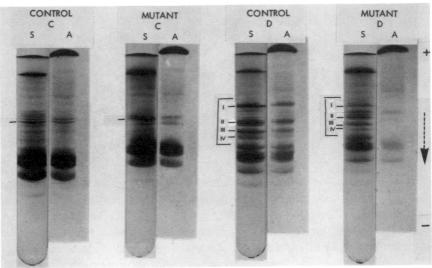

Figure 4.16. Electrophoretograms on acrylamide gels, of proteins of *Xenopus* embryos. After Hallberg and Brown, 1969. Four embryonic regions: A, B, C and D are compared between normal and anucleolate mutants. In each pair of tubes, S is the stained gel and A the autoradiogram obtained after labelling with radioactive precursors to detect newly-synthesized proteins. Note that the ribosomal proteins, bands I–IV, hardly show at all in the autoradiograms from mutants, showing that little or no synthesis of these proteins occurs in them

(d) DNA Synthesis and Storage in Oocytes

Several features about the types of DNA synthesized in oocytes of *Xenopus* and their location have been reviewed by Brachet and Malpoix (1971). They affirm that at the pachytene stage of the first meiosis, two distinct types of DNA can be recognized in the nucleus of the primary oocyte. These are: chromosomal DNA, and 'nucleolar organizer' DNA (i.e. rDNA). The extrachromosomal, rDNA is of particular interest. Coggins (1973) has shown recently by labelling with radioactive thymidine that this rDNA begins to be synthesized as a 'halo' round the already numerous nucleoli, at the zygotene stage of the first meiosis. Later, it accumulates as a 'cap' of extrachromosomal DNA at one end of the nucleus. It has been shown by molecular hybridization *in situ* that this cap material hybridizes with rRNA, so there is no doubt that it contains rDNA. Birnstiel, Grunstein, Speirs and Hennig (1969) concluded that it contains 800 copies of a base sequence that is complementary to rRNA. So it is clear that this DNA could provide sufficient rRNA by transcription, to build up the large reserve of ribosomes that accumulates in the oocyte. Moreover, the DNA has been shown to enter the oocyte nucleoli when these form: MacGregor (1968) described the passage of the cap material first into granules (micronucleoli, perhaps) in the nucleoplasm and then into the cores of the nucleoli.

As we saw earlier (section (c)), the oocyte nucleoli all disappear at the second meiotic division. Where they go to is uncertain: but Brachet (1968) reported an accumulation of Feulgen-positive (i.e. DNA-containing) particles at the animal pole of the meiotic spindle, which passed into the egg cortex and then vanished (Brachet, Hanocq & Van Gansen, 1970). It is interesting that at just about this time, germ plasm, rich in RNA, begins to be detectable in the inner zone of the egg cytoplasm (Czołowska, 1969: cf. section 3). So Wallace, Morray and Langridge (1971) have suggested that some of the rDNA from the nucleoli may become transcribed as RNA and pass into this germ plasm, since it appears at just the time when the nucleoli disappear. If so, the germ plasm does not represent any special store of genetic information by which species characters might be inherited through the germ line, but is simply a store of extra rRNA to provide for ribosomes of the next generation. The disappearance of the germ plasm in stage 44 larvae also fits logically into this assumption, for by that stage there should be plenty of new ribosomes being formed under the influence of the somatic cell nucleoli. One could still perhaps argue that some of the maternal rRNA in the germ plasma may form special ribosomes which are sites for synthesis of special proteins of maternal type in the early embryo, and that thus some special characters can be 'inherited' via the germ plasm. But to support this whole idea that the germ plasm is derived from transcription of nucleolar DNA, we need evidence that the RNA seen in the early germ plasma is rRNA. So far, this point has not been established.

Another possibility might be that the DNA of the disappearing nucleoli becomes incorporated into cytoplasmic reserves of DNA in the egg. Woodland (1969) observed an increase in thymidine phosphorylation at this stage when the nucleoli disappear in *Xenopus*, which suggests a large-scale synthesis of DNA in the cytoplasm. The yolk platelets are thought to be the main site of storage of cytoplasmic DNA in amphibian oocytes, from histochemical studies (Baltus & Brachet, 1962). Mitochondria also contain DNA, but this has a special molecular structure (Dawid & Wolstenholme, 1967) and is unlikely to be a reserve for future nuclear synthesis. During cleavage of the embryonic nuclei (see Chapter 6) there is very little overall increase of the DNA per embryo, which indicates that the cytoplasmic reserves are used for nuclear synthesis.

(e) Types of RNA Synthesized during Oogenesis

The course of RNA synthesis in oocytes of *Xenopus* has been followed both qualitatively and quantitatively, by autoradiographic and biochemical methods. Ficq (1961) showed that labelled cytidine and uridine are taken up preferentially into the nucleoli, when these are present: so evidently these become the most active sites of RNA synthesis in the oocyte. The types of RNA synthesized have been studied by Ford (1970). He found that small oocytes have relatively little rRNA, but that there is a 42s category of RNA which may be the precursor of rRNA. During early to mid-oogenesis (stages B and C of Balinsky and Devis's descriptions, see section 2), mainly transfer RNA (tRNA) is synthesized, and this eventually makes up 5–10 % of the total RNA. Heterogeneous, DNA-like RNA (i.e. dRNA) is also formed at these stages and makes up 10–20 % of the total RNA. There is also some synthesis of 5s RNA, which is the smallest molecular size that contributes to ribosomes. But the synthesis of the other two ribosomal types, 18s and 28s RNA, lags behind that of 5s, which is surprising since ribosomes eventually have a 1 : 1 : 1 ratio of the three molecular types. Ford (1971) drew attention to this early excess of 5s RNA, and suggested that it was a reserve to compensate for the fact that there are many more DNA sequences that code for 18s and 28s RNA than for 5s RNA in *Xenopus*. Must one then assume that later, the activity of the different DNA types is regulated so as to produce the necessary 1 : 1 : 1 ratio of 5s : 18s : 28s RNA when ribosomes start to form? This assumption may be unnecessary, in fact, for it has been realized for some time that not all genes in a genome are necessarily active and that there may be much redundancy: so the 18s and 28s RNA may not be produced in proportion to the numbers of gene copies present. Recently Amaldi *et al.* (1973) have produced evidence and arguments suggesting that many of the genes in *Xenopus* are 'redundant' for somatic cells, and function only in the oocyte to produce extra low-molecular-weight RNA, nucleolar RNA and histones. The same redundancy argument may apply to some of the genes in the oocyte: particularly those that already have multiple copies.

The gross reserves of RNA in the cytoplasm of the oocyte are, like DNA, apparently bound to the yolk platelets (Kelley, Nakai & Guganig, 1971). It is also possible that some forms of RNA are synthesized in the oocyte cytoplasm (Woodland & Gurdon, 1968b). Rosbash and Ford (1974) have observed mRNA here, and Kessel (1969) claimed to have seen nucleolus-like bodies in association with the mitochondria, which might be capable of synthesizing rRNA. Swanson and Dawid (1970) have isolated a ribosome-like particle from mitochondria of the oocyte, and have shown that it is capable of protein synthesis. Roeder (1974) has also observed DNA-dependent RNA polymerase activity in oocytes of *Xenopus*. The degree of autonomy in developmental processes that is possible in the cytoplasm of amphibian oocytes is emphasized also by work of Smith and Ecker (1969) on *Rana*. They showed that the maturation of the oocyte occurs in response to gonadotrophins *in vitro*, and that most of the processes involved can be brought about equally well in oocytes whose nuclei have first been removed surgically. Special properties of oocyte and egg cytoplasm have also been highlighted in the nuclear transplantation work of Gurdon and his collaborators, discussed in Chapter 6. They found that oocyte cytoplasm stimulates RNA synthesis, but inhibits DNA synthesis, in transplanted brain nuclei. Egg cytoplasm has just the opposite effect, inhibiting RNA synthesis but stimulating DNA synthesis (Graham, Arms & Gurdon, 1966; Gurdon, 1967).

5. The Initiation of Development in the Egg, by Fertilization or by Parthenogenesis

Before its conversion into an egg occurs, the oocyte must complete meiosis and be shed from the follicle in the ovary. The stimulus for this event, as we saw in the last chapter, is the hormone LH from the anterior pituitary, which will also act on oocytes *in vitro*. Thornton (1971a) used this *in vitro* test as a method of assaying gonadotrophic hormones.

After the first meiotic division is complete, further events in the shed oocyte are normally initiated by fertilization. In *Xenopus* as in other anurans, the secondary oocyte is laid and fertilization occurs externally. In response to the entry of a spermatozoön the secondary oocyte undergoes a further meiotic division, giving off a small second polar body which contains one haploid set of chromosomes, and itself becoming the egg with a haploid pronucleus. (cf. Figure 4.1 above). The female pronucleus then fuses with the haploid pronucleus of the spermatozoön. The normal process of fertilization in Amphibia has been described in many text books, so only a few features special to *Xenopus* will be mentioned below.

Balinsky (1966) and Grey *et al.* (1974) have followed in ultrastructural studies the changes that occur after fertilization in *Xenopus*, at intervals from 30 sec to 1 hr after the entry of the spermatozoön. Balinsky observed that, first, the cortical granules which have been present in the egg cortex from oocyte stages (cf. section 4) discharge their contents into the perivitelline space, between the egg surface and its vitelline membrane (Figure 4.17). This occurs at $2\frac{1}{2}$–$3\frac{1}{2}$ mins after sperm entry. Next, at 10 min, a honeycomb of microvilli appears over the surface of the animal hemisphere of the egg. These microvilli are only transitory, however: during the next half hour they flatten out, and at the same time the egg cortex is displaced in such a way that in a

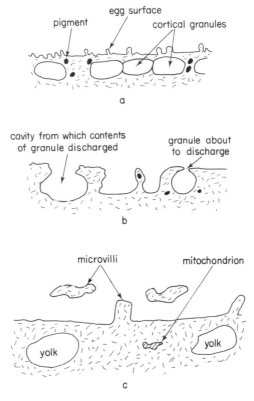

Figure 4.17. Discharge of cortical granules into the perivitelline space at fertilization: seen with the electron microscope. Redrawn after Balinsky, 1966. (a) Before fertilization; (b) $2\frac{1}{2}$ min. after fertilization; (c) 3 min. after fertilization: all granules discharged

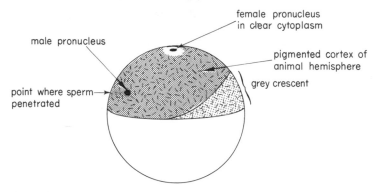

Figure 4.18. Diagram to show the position of the grey crescent in the amphibian egg after fertilization

crescentic zone diametrically opposite to the point where the sperm entered, the cortical pigment is dispersed. The pale crescent that results is known as the 'grey crescent' (Figure 4.18). It is seen in other amphibian eggs too at the initiation of development, and marks the future dorsal side of the embryo. Towards the end of the first hour after fertilization in *Xenopus*, changes also occur in the subcortical layers of the egg. Balinsky observed that the subcortical cytoplasm increased in volume, by dispersal of vesicles (Figure 4.19). An agglutination reaction ensues (Wyrick *et al.* 1974), preventing entry of further sperm.

Besides studies on normal fertilization, there have been a number of attempts to initiate development in the egg of *Xenopus* by artificial means. This may result in embryos with abnormal numbers of chromosomes, for the final meiotic divisions may fail or may take place abnormally in these cases. The diploid chromosome number in *Xenopus l. laevis* is 36 (Wickbom, 1945). Androgenetic haploids, in which only the 18 chromosomes from the spermatozoon are present, are relatively easily produced in *Xenopus*, since

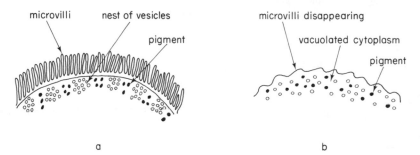

Figure 4.19. Diagrams showing the contribution of nests of vesicles to the subcortical cytoplasm in *Xenopus* eggs. Redrawn after Balinsky, 1966. (a) A few min. after fertilization. (b) 1 hr. after fertilization

the female pronucleus is easily killed by irradiating the upper pole of the egg with a specific dosage of ultraviolet light (Gurdon, 1960b). In other species it can only be eliminated by removing it surgically, which is more difficult. Gynogenetic haploids, in which only the egg nucleus is represented, can also be produced in *Xenopus*, by irradiating the sperm to inactivate their nuclei but then allowing them to penetrate the egg and so initiate its development (Selman, 1958). These possibilities are shown diagrammatically in Figure 4.20. Haploid gynogenetic development will also occur if mature

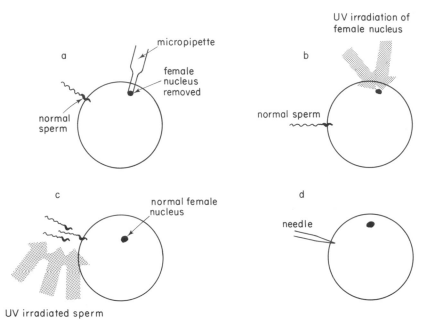

Figure 4.20. Diagrams to illustrate possible methods of producing haploids and parthenogenetic development in amphibians. (a) Androgenetic haploid (*Rana*). (b) Androgenetic haploid (*Xenopus*). (c) Gynogenetic haploids by irradiating sperm. (d) Gynogenetic haploids by pricking

oocytes are activated simply by pricking them: this is most effective if the needle is contaminated with serum (Rostand, 1951). The percentage success is very low, however: Rostand obtained only 20 larvae from a total of 2,000 eggs, and these larvae survived for only 9 days. They showed the usual 'haploid syndrome' described by Hamilton (1963), which includes stunting of the body and oedema. Haploids are also infertile.

Triploids have also been obtained in *Xenopus*, with a higher proportion of success than with haploids. Smith (1958) found that a 5-min heat shock (36·1–36·5 °C) within 10 min after fertilization suppressed the formation of the second polar body, so that the chromosome number doubled in the egg, and triploids resulted, in 54 % of cases. The triploids developed to maturity, but their fertility was low, probably owing to failures of meiotic divisions during gametogenesis. Tetraploids have also been obtained by various means: Gurdon (1959) observed a very low proportion of tetraploids among embryos resulting from nuclear transplantation experiments, where the transplanted nucleus had evidently undergone an abnormal first division leading to doubling of its chromosome number. He also quotes other methods that have occasionally produced tetraploids: viz. cold shock treatment of fertilized eggs resulting in fusion of the first two cleavage nuclei, and matings between diploids and triploids. Spontaneous tetraploidy has also been reported in *Xenopus* by Fischberg, Elsdale and Smith (unpublished). Tymowska and Fischberg (1973) find *X. l. bunyoniensis* 'tetraploid' compared with *X. l. laevis*.

Gurdon made some interesting observations on the behaviour and gonadal anatomy of the five male and five female tetraploids that he was able to rear to maturity from nuclear-transplant eggs. The males were less able to croak and to clasp females than were diploids, and their testes were reduced in size. Plenty of spermatogonia and primary spermatocytes were present, but these failed to undergo normal meiosis, so that the nuclei of the spermatids were pycnotic and the spermatozoa misshapen. Tetraploid females were not clasped by normal males, which suggests that they did not show the normal behaviour pattern. Their oogenesis also failed at meiotic stages.

Although these animals with unusual chromosome complements are obtainable only in small numbers and are unlikely to be fertile, they are of potential use in genetic studies, as are the anucleolate mutants. So far we have very little information on the genetics of *Xenopus laevis*, although the work described in this chapter, on the germ plasm and on nucleolar DNA in oocytes, has laid open a wide range of possibilities for future genetical work. Mutations, and hence characters controlled by single genes, are far easier to discover in animals which show them as coat-colour variations, as do mice, than they are in the amphibians like *Xenopus* which have no very easily-distinguished external features controlled by known single genes. The great advantage of the amphibians, however, is the accessibility of their eggs and embryos for experimental work. Thus we know far more about the embryonic development of *Xenopus* than we do of its genetics. We must now go on in the following chapters to consider some of these findings on the developmental biology of *Xenopus*.

5

Development of the Embryo and Larva of *Xenopus laevis*

In later chapters we shall be referring to several events that occur during embryonic and larval stages in *Xenopus*. This chapter will therefore give a short introductory description of embryonic and larval development, emphasizing the external features by which each stage is recognizable, as well as explaining briefly the internal morphology and how this develops. Many readers will already be familiar with the development of the common frog, *Rana* which is a standard text-book 'type' in biology teaching. *Xenopus* differs only a little from this 'type', and we shall point out these differences as we go along. The development of the embryo from fertilization to hatching will be described first, then the development of certain features in the larva up to and including its metamorphosis. For fuller details of the development of *Xenopus laevis*, the standard work is Nieuwkoop and Faber's Normal Table (1956), which gives many references to the original descriptions and is an invaluable handbook for anyone working on this species.

1. Development of the Embryo

As in descriptions of other vertebrate embryos, certain phases of development in *Xenopus* have been given conventional names. First comes *cleavage*, when the zygote (fertilized egg) undergoes a series of five or six synchronous divisions into 2, 4, 8, 16 etc. cells or *blastomeres* (Figure 5.1. (a)–(d)). This is followed by a less regular series of divisions during which the originally solid ball of cells (*morula*) acquires an internal cavity and is then called a *blastula* (Figure 5.1 (e)). Next comes the phase known as *gastrulation*, when a proportion of the cells gradually migrate below the surface to form two inner layers, the *mesoderm* and *endoderm*, while at the same time the cells which remain outside spread to cover the whole surface of the embryo, forming the *ectoderm*. The invagination process begins at a dorsally placed groove known as the *blastopore* (Figure 5.1. (f)) and gradually spreads

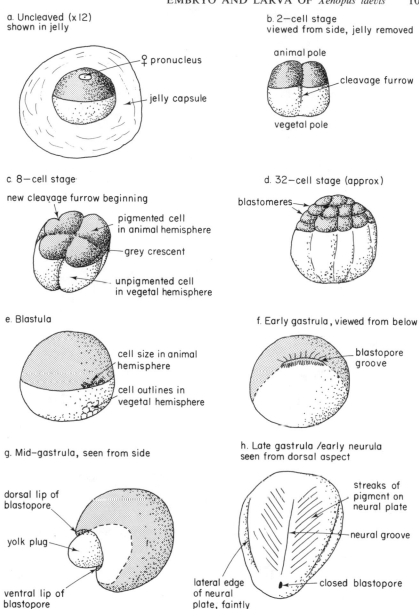

a. Uncleaved (x 12)
shown in jelly

♀ pronucleus

jelly capsule

b. 2-cell stage
viewed from side, jelly removed

animal pole

cleavage furrow

vegetal pole

c. 8-cell stage

new cleavage furrow beginning

pigmented cell
in animal hemisphere

grey crescent

unpigmented cell
in vegetal hemisphere

d. 32-cell stage (approx)

blastomeres

e. Blastula

cell size in animal
hemisphere

cell outlines in
vegetal hemisphere

f. Early gastrula, viewed from below

blastopore
groove

g. Mid-gastrula, seen from side

dorsal lip of
blastopore

yolk plug

ventral lip of
blastopore

h. Late gastrula /early neurula
seen from dorsal aspect

streaks of
pigment on
neural plate

neural groove

lateral edge
of neural
plate, faintly
indicated

closed blastopore

Figure 5.1 (and 5.2). External views of *Xenopus* embryos, from cleavage to hatching
stages
Figure 5.1. (a) uncleaved egg, (b) 2-cell stage, (c) 8-cell stage, (d) 32-cell stage, (e)
blastula, (f) early gastrula, (g) mid-gastrula, (h) late gastrula. *Figure 5.2.* (i) early
neurula, (j) late neurula, (k) tailbud stage, (l) hatched larva

i. Early to mid−neurula (x 20)

j. Late neurula, viewed from anterior end

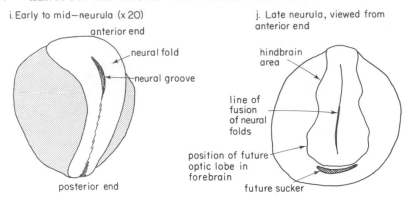

k. Tailbud stage : lateral view−(x 15)

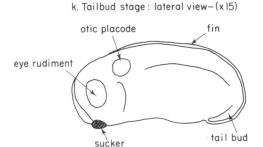

l. Hatched larva : lateral view (x 10)

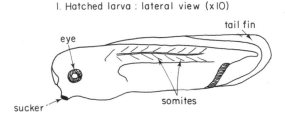

Figure 5.2. External views of *Xenopus* embryos (continued)

laterally and ventrally. As it does so, the blastopore groove becomes circular and surrounds a 'yolk plug' of endoderm cells (Figure 5.1 (g)). Towards the end of gastrulation, when invagination is finishing and the last endoderm cells are finally drawn in, the diameter of this circular blastopore shrinks, until it and the yolk plug disappear altogether (Figures 5.1 (h)–5.2 (i)). There then follows the phase known as *neurulation*, when the future nervous system appears as a thickened *neural plate* in the dorsal ectoderm, and this plate rolls up at its sides to form a tube (Figure 5.2 (j)). At the end of neurulation, the margins of the tube have fused, and the regions representing future brain and spinal cord are distinguishable (Figures 5.2 (j)–(l)). After this there is little external change until eye rudiments, a sucker and a tail bud appear (*tailbud* stage). A fin forms round this tail bud, and muscular movements begin which cause the embryo to break out of its jelly capsule and become a hatched larva. This whole process, from fertilization to hatching, is more rapid in *Xenopus* than in most European amphibians, taking only 72 hrs at 18 °C. We shall now consider each phase in a little more detail.

(a) Cleavage

This process is fascinating to watch under a low-power dissecting binocular. The first two cleavages are especially easy to see from above, since they are meridional and start at the upper, 'animal' pole of the zygote. Here the nuclei lie quite close to the surface in a small, circular area of clear cytoplasm, so they can be seen to divide first, then a series of wrinkles and eventually a rapidly-spreading and deepening furrow can be seen to form (Figure 5.3). Pigment streaks towards this furrow, but usually does not extend right down into it, so that its depth is white, which makes it all the more easily visible. At first the furrow does not extend right to the ventral pole of the embryo, nor does it extend right through from side to side, so the daughter cells have a broad, open zone internally where their cytoplasm is still confluent (Figure 5.4). A thin membrane is gradually formed internally, closing this region so that the two cells are eventually separate, but the first internal cell membrane is not complete until two or three more cleavages have taken place.

After the first two cleavages, the next one is normally horizontal, cutting off four smaller cells at the animal pole and four larger cells at the vegetal pole (Figure 5.1 (c)). There is some variability in the cleavage pattern in individual *Xenopus* embryos, however, and odd numbers of cells may occasionally be seen, caused by slightly asynchronous divisions. Apart from these aberrant cases, it is normal for cleavages to continue synchronously up to at least the 64-cell stage, by which time the cells are no longer tightly packed at their internal surfaces, but are beginning to surround an internal cavity, the blastocoel (Figure 5.5): so the embryo is now a blastula. The cells of the upper, pigmented animal hemisphere are considerably smaller than those of the lower, paler vegetal hemisphere. This difference in size becomes accentuated as synchrony breaks down and the animal pole cells divide more rapidly than the vegetal ones. By the time gastrulation is due to begin, it is barely possible to distinguish individual cells in the animal hemisphere with the naked eye, because they are very small (Figure 5.1 (e)).

(b) Gastrulation

The course of gastrulation in amphibians has been described very clearly in text books of embryology, so we shall concentrate here only on certain features that are special to *Xenopus*. The first unusual feature, observed by

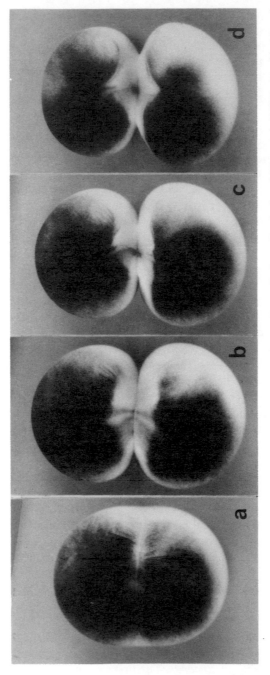

Figure 5.3. External views of the first cleavage in *Xenopus* embryos. From DeLaat, Luchtel and Bluemink (1973). (a), (b), (c) and (d) show the progress of the cleavage furrow at 5, 14, 20 and 24 min. respectively after the appearance of a pigment stripe where the furrow will form. This stripe also marks the beginning of anaphase in the nuclear division

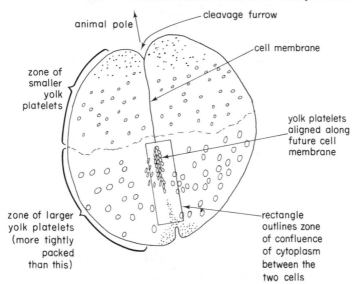

Figure 5.4. Histological section through 2-cell stage embryo. Linear magnification × 48

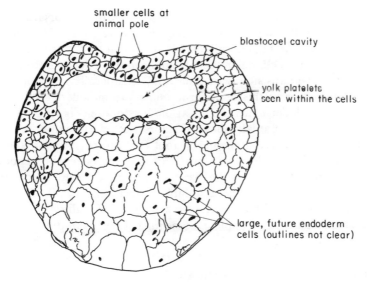

Figure 5.5. Transverse section through early blastula. Linear magnification × 50

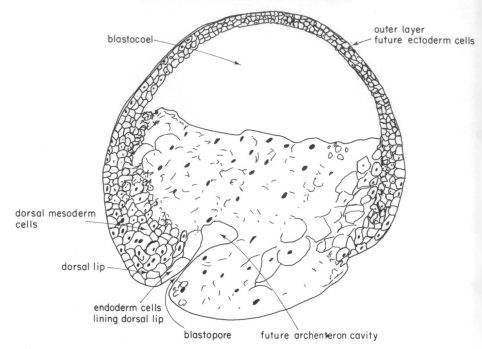

blastocoel

outer layer
future ectoderm cells

dorsal mesoderm
cells

dorsal lip

endoderm cells
lining dorsal lip

blastopore   future archenteron cavity

Figure 5.6(a). Longitudinal section through early gastrula. Linear magnification × 60

Nieuwkoop and Florschütz (1950) is that some of the future dorsal mesoderm is already below the surface before the blastopore forms. It is not fully understood how this mesoderm reaches its position, but the most likely possibility is that the cells delaminate from the inner surfaces of the cells that lie above them, perhaps simply by tangential divisions. The blastopore then forms just posterior to this mesoderm, and more mesoderm, as well as a lining layer of endoderm, invaginate at the blastopore (Figure 5.6 (a)). All these tissues migrate cranially under the dorsal ectoderm, until they form a complete, double-layered archenteron roof (the future gut roof, shown in Figure 5.6 (b). Meanwhile, some invagination of mesoderm and endoderm begins to occur both laterally and ventrally, round the lateral and ventral lips of the blastopore, as in other amphibians (cf. Figure 5.7 (a)). In this way the future side walls and floor of the gut are formed. Finally, the blastopore closes and gastrulation is complete. Some further details of the invaginating cells are shown in Figure 5.7 (b). A noteworthy point is that they are not flask-shaped, as described in Holtfreter's classic account (1944) of gastrulation in the newt, but are of varying shapes and sizes.

At the end of gastrulation it is very difficult externally to distinguish the dorsal from the ventral side of the embryo in *Xenopus*. The dorsal surface is

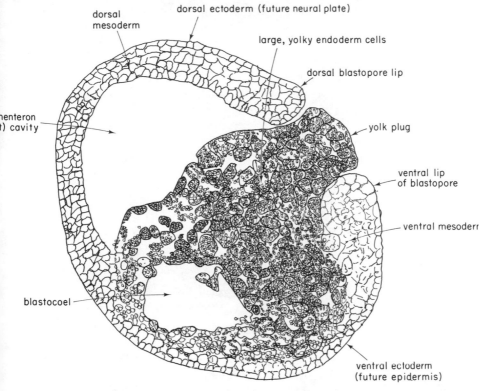

Figure 5.6(b). Longitudinal section through late gastrula. Linear magnification × 60

slightly paler, however, and in an embryo whose membranes have been removed, it may appear flattened. One can, of course, also distinguish dorsal from ventral by dissection, since the roof of the gut is much thinner than its floor (Figure 5.6 (b)).

The above description is based purely on observations that can be made with the dissecting binocular and by light microscopy. Recently Tarin (1971a) has re-examined the external features of *Xenopus* embryos by scanning electron microscopy, and some of his pictures show surface details very clearly. They do not, however, provide new information about the mechanism of gastrulation, so have not been included here.

(c) Neurulation

This is not so clearly visible externally in *Xenopus* as it is in urodeles. Anyone wishing to demonstrate neurulation to students is advised to use newt or axolotl embryos, for even those of *Rana* do not have a very obvious neural plate: nor does *Xenopus*. At first one can just make out streaks of

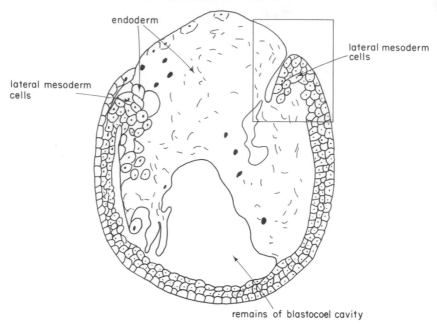

Figure 5.7(a). Horizontal section through early gastrula. Linear magnification × 60

pigment along the future midline neural groove and the lateral edges of the neural plate (Figure 5.2 (i)). Next, the lateral edges come closer together, but do not rise up obviously as folds, like they do in urodeles. Later, the margins of the neural plate do become more prominent however, as they meet and fuse in the mid-line (Figure 5.2 (j)). This fusion begins in the cervical region and proceeds cranially and caudally from here. By the time the neural folds have fused completely at the cranial end, the contours of the future brain are evident, including the eye rudiments (Figure 5.2. (k)).

Internally, there are several changes in the mesoderm during neurulation. The dorsal mesoderm, lying on the roof of the gut and in contact with the ectoderm of the future neural plate, is in fact responsible for *inducing* the neural plate to form (see Chapter 7). It also becomes organized at this time into a mid-line rod, the notochord, which will later have cartilage of the future vertebrae deposited round it, and a series of segmented blocks of mesoderm on either side of the notochord, called the somites (Figure 5.8) from which vertebrae, segmental muscles and dermis of the skin are derived.

There is little morphological change in the endoderm during neurulation, but it elongates slightly and becomes more compact, with a surface epithelium of smaller cells covering the large, yolky cells which still cause the floor of the gut to be thick. Among these large cells are the germ cells, which as we saw in Chapter 4 will later migrate dorsally into the future gonads.

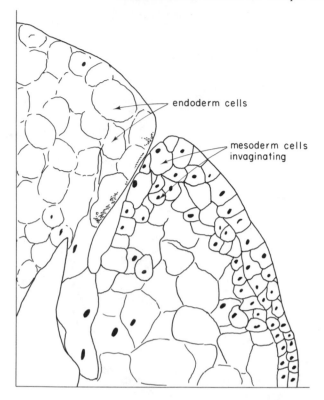

endoderm cells

mesoderm cells
invaginating

Figure 5.7(b). Detail of invaginating cells in Figure 5.7(a). Linear magnification × 220

Externally, the epidermis forms from that part of the ectoderm which does not form neural plate, and it becomes ciliated during neurulation. The cilia beat in an anterior-to-posterior direction, so that if the embryo is divested of its jelly and membranes, it will move slowly forwards, though there is also a tendency for it to rotate, owing probably to the stronger beating of dorsal cilia than ventral ones.

(d) Tailbud to Hatching

The segmentation of somites and the axial organization of the mesoderm continue after the end of neurulation, and the somites can be seen externally owing to the development of grooves between them (Figure 5.2 (l)). They are visible in longitudinal section too (Figure 5.9). The eye rudiments become more prominent and the head takes on a distinctive, triangular shape as viewed from the cranial end. At the future mouth opening there is already at this stage a darkly-pigmented mucous gland. This secretes a mucous thread after hatching, which serves to suspend the larva from the surface of the

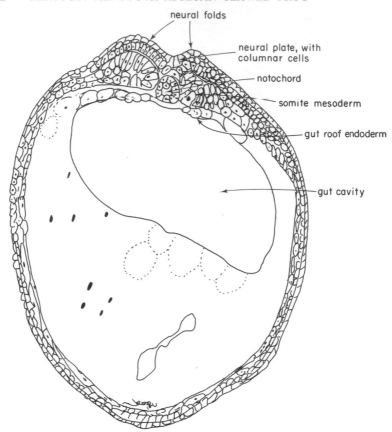

Figure 5.8(a). Transverse section of early neurula. Linear magnification × 80

water, and later to draw food particles into its gullet. It is also thought to contain enzymes which facilitate the hatching process by dissolving the vitelline membrane and jelly. At the caudal end, the tail bud appears as a small knob, pointing ventralwards at first, but this later assumes a horizontal position as it elongates (Figure 5.10 (a), (b)). In *Xenopus*, unlike some other amphibians, the blastopore closes in such a way as to leave much of the future tailbud mesoderm ventral to it: then a new opening develops further ventrally, below the tailbud mesoderm, to form the anus and this is not derived from the blastopore as it is in some other anuran species (cf. Bijtel, 1931).

Other external features that appear before hatching are the fin (Figure 5.2 (l)), developed from superficial epidermis and mesoderm around the tailbud, and a slight swelling on each side in the region of the future gills which are developing from pharyngeal endoderm and mesoderm below the

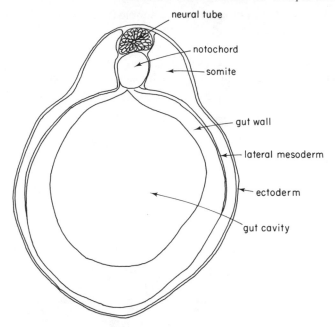

Figure 5.8(b). Transverse section of late neurula, after folds have closed. Linear magnification × 60. Diagrammatic, as cell outlines are not clear at this stage in histological section

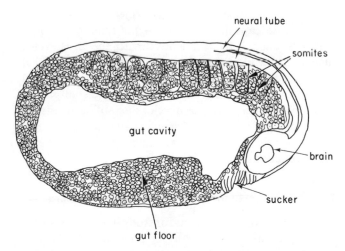

Figure 5.9. Longitudinal section of tailbud stage, a little lateral to mid-line so as to show somites. Linear magnification × 40

surface. The rest of the gut remains as a simple tube, but now has an anus, a liver diverticulum and a slightly more tubular foregut region behind the pharynx, where the oesophagus and stomach will form. The whole animal elongates to about three times the original length of the neurula, and so has to bend laterally in order to fit into the jelly capsule. This situation seems to

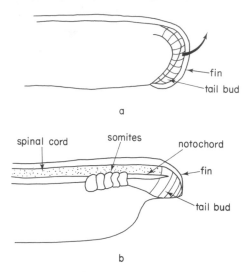

Figure 5.10. Longitudinal section of tailbud stages: diagrammatic. (a) early stage; (b) later stage

initiate vigorous side-to-side bending movements, by which it eventually breaks out from the capsule, tail first. After remaining for some hours attached to its jelly capsule by its mucous thread, the larva eventually swims away and uses this thread to suspend itself vertically from the surface of the water. From this complex hatching behaviour it is clear that elaborate neuromuscular functions are already present. As we shall see in Chapter 9, there are many special features in the development of both sensory and motor elements in the nervous system of *Xenopus*. Synapses have been seen in the walls of the neural tube, in late neurula stages (Roberts and Hayes (1973)), so some neural functions may begin as early as this.

2. Development of the Larva up to the Time of Metamorphosis

Xenopus larvae are attractive objects for study in the laboratory, as they are remarkably transparent, so that many of their internal organs are visible *in vivo*. They have extraordinarily large, transparent heads in which the optic stalks and the pharyngeal region, including the thymus glands, can be seen very clearly. A number of observers have described their external features, and the brief account given below will be based partly on descriptions by Weisz (1945 (a), (b)) and partly on Nieuwkoop and Faber's Normal Table (1956) as well as on personal observations. We shall deal first with the external features and then refer to a few points of interest in the organ systems. More will be said about experimental work on the development of organ systems, in Chapter 9.

(a) External Features

The newly-hatched larva of *Xenopus* (Figure 5.2 (l)) is about 4 mm long and 1 mm wide, its maximum width being at the level of the eyes. Its ventral regions appear yellowish owing to the yolk which remains in the endoderm cells of the gut. The dorsal colouring of the larva is grey to transparent, except for the eyes which already contain melanin and therefore show up black through the skin. The pre-oral sucker at the ventral tip of the head is also black. Above the sucker is a frontal gland which also secretes mucus and helps to suspend the larva from the surface of the water.

For the first 16–24 hours after hatching, the larva remains fairly quiescent, either on the bottom of the aquarium or suspended from the water's surface by its mucus thread. During this time the mouth is developing, however, and when this has opened through to the pharynx so that feeding is possible, the larva begins to swim actively and to ingest material that has become trapped in the mucus thread, which is now drawn into the mouth and eventually down into the stomach (Dodd, 1950). Vestiges of the sucker remain until six days after hatching, when they disappear completely.

During the period from the 2nd to the 7th day after hatching (Stages 45–49 of Nieuwkoop & Faber's Table), the larva is referred to by Weisz as the 'first-form tadpole'. During this phase, two pairs of transitory external gills develop. At the end of this period the most remarkable change is the disproportionate growth of the head: the ratio of the lengths of head and abdomen is 1:1 at first but has become 2:1 by the 7th day. More skin pigmentation also develops during this period.

From the beginning of active feeding on the 7th day after hatching until the appearance of the hind limb buds at about 35 days (Nieuwkoop and Faber's stages 49–56), Weisz referred to the *Xenopus* larva as the 'second-form tadpole' (Figure 5.11). The chief external change during this period is a gradual acquisition of more cutaneous pigment, tending to obscure the internal organs from the dorsal aspect, though they are still visible from below. The head does not show any further increase in size relative to the body, though both are growing steadily and the head becomes broader and squarer in shape. The characteristic stance of the larva at this stage is in an oblique position with the head slightly lower than the tail, and the tail tip constantly vibrating from side to side to help maintain the animal in this

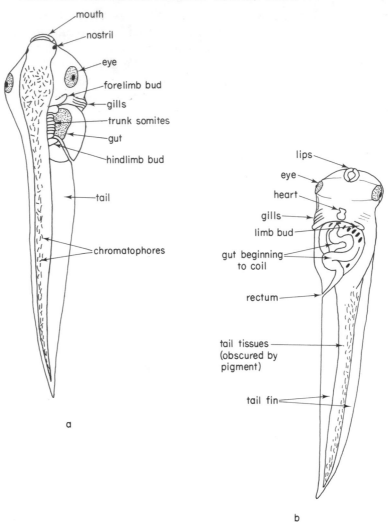

Figure 5.11. Second-stage larva of *Xenopus*: (a) dorsal, (b) ventral. Structures seen by transparency in living specimen under MS 222 anaesthesia

position. While remaining stationary, it carries out continuous gulping movements, taking in water at the mouth and passing this out over the gills, which are now internal. Food materials contained in the water are trapped by a system of folds, cilia and mucus in the pharynx and carried down to the stomach by ciliary action.

From the 5th to the 10th week after hatching (Nieuwkoop and Faber's stages 56–60 plus) is Weisz's 'third-form tadpole' stage. During this phase, the main change is further growth, and the larva's length increases from 30 mm to 50–80 mm. The head remains about twice as large as the abdomen, the relative measurements at 10 weeks being: length of head 7 mm, abdomen 3 mm and tail 20 mm; width of head 6 mm and of abdomen 3 mm. The hind limbs also grow rapidly during this period, and towards the end of it, forelimb buds emerge and metamorphosis begins. Metamorphosis involves radical changes in body form, which will be described in section 3.

(b) Central Nervous System

The morphology of the developing brain and spinal cord in larval *Xenopus* does not differ noticeably from that in other Anurans. The brain of the larva has large olfactory lobes, small but prominent optic lobes and a very small cerebrum area. It develops from the three vesicles—forebrain, midbrain and hindbrain—which are already evident at the tailbud stage (cf. section 1). At hatching the forebrain is still bent ventrally, but as the larval body straightens and elongates, this brain flexure is reduced until the forebrain comes into alignment by $1\frac{1}{2}$ days after hatching (stage 43). The pineal gland and pituitary are already formed by this stage, and all intracellular yolk has been absorbed. Further details of brain structure are laid down during the next three weeks, and the brain is fully formed by stage 53 (22 days after hatching).

In the early larval spinal cord, an interesting feature is the series of segmentally-arranged dorsolateral sensory cells, called Rohon-Beard cells (Figure 5.12) which are also found in other amphibians. They are present in *Xenopus* from hatching until stage 50 (13 days after hatching) when they begin to degenerate and are replaced by spinal ganglia. Sympathetic ganglia

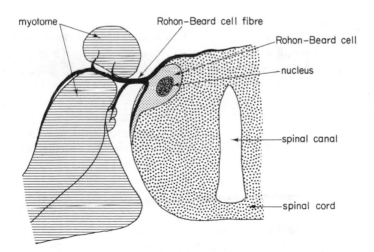

Figure 5.12. Diagram to show a Rohon–Beard cell in the spinal cord of an amphibian larva. After Coghill, 1914

are also present by stage 50. The spinal cord and its 9 pairs of trunk ganglia are complete, with all fibre tracts present, by stage 66, when metamorphosis has been completed. There are prominent brachial and lumbar enlargements in the cord, caused by the more numerous motor neurones at limb levels: a feature which has been noted in other amphibians as well as in birds and mammals (cf. Chapter 9).

(c) Skeleton

The skeleton and muscles of the developing larva of *Xenopus* have been described by Weisz (1945a) in some detail. The pharyngeal cartilages develop earliest, chondroblasts accumulating in this region in the first-form larval stage. The cartilaginous skull (chondrocranium) also begins to form at this time, and it and the axial skeleton are laid down during the second-form larval stage. The limb girdles do not appear until the third-form larva, and there are no limb cartilages until near the beginning of metamorphosis. Rather surprisingly, the transverse processes of the vertebrae are not fully formed until metamorphosis, either. Nieuwkoop and Faber (1956) give an interesting summary description of the gradual appearance of plates of mesenchyme which give rise to the different cartilaginous elements of the skeleton during larval life. The first of these to appear is the mesenchymal rudiment of the ethmoid plate in the floor of the chondrocranium. A general point of interest is that all of the axial skeleton develops from paired rudiments, which gradually approach each other and fuse in the mid-line. Major changes in the form of the skeleton, as well as its ossification, take place during metamorphosis (see section 3).

(d) Vascular System

The vascular system of *Xenopus* larvae has been of great interest to comparative anatomists owing to the absence of an efferent branchial artery system (cf. Chapter 2), unlike other vertebrates with gills. This may indicate that the gills in *Xenopus* are not essential in respiration, though as we saw earlier, there is still some disagreement on this. As development proceeds in the larva, there is a tendency for the aortic branches supplying the gills to decrease, though those with branches to the filter apparatus remain (cf. Figure 2.28, p. 54). Some of the main changes during development of the vascular system, which have already been described in Chapter 2, are clearly visible *in vitro* in the larva because of its transparency. At hatching, it can be seen that the heart is still S-shaped and that the ventral blood vessels are ill-defined as they are not yet lined with endothelium. The aortic arches are present already, and the pulmonary arteries and veins soon appear, since the lungs develop very early in larval life. In the first-form larva, the heart is seen to have two auricles, the first aortic arch has dwindled and arches 5 and 6 now supply the external gills. The originally paired dorsal aortae are already fusing to form a single median vessel in the trunk. The venous system also quickly undergoes major changes, jugular veins replacing the anterior cardinals and the inferior vena cava soon superseding the posterior cardinal veins.

By the second-form tadpole stage, the external gills have disappeared and so has the dorsal part of aortic arch 5. This situation persists in the third-form larva, which also loses the ductus arteriosus. During this third-form phase, the truncus leading from the heart becomes subdivided into aortic and pulmonary roots by the development of a spiral valve. Lymph spaces also begin to appear, and these become very extensive under the skin after metamorphosis.

Further landmarks in the development of the vascular system in *Xenopus* are, according to Nieuwkoop and Faber's account: the appearance of the median abdominal vein draining the ventral body wall by stage 42 ($1\frac{1}{2}$ days after hatching); the entry of the pulmonary vein into the right atrium by stage 44 (2 days after hatching) and the appearance of limb arteries and veins in the early second-stage larva at stage 50–51 (13–15 days after hatching). At this time both fore and hind limb buds are still very small.

(e) Pharynx and Gut

Much interest has been centred on the pharynx in *Xenopus* larvae, since it includes a special filter-feeding system and is remarkably large. Even at hatching, the pharynx is already wide and the pouches representing future gills are present. In the first-form larva, further lateral expansion of the pharynx occurs and its lining acquires bands of cilia, extending into each gill pouch. There is now a continuous ciliated channel from the pharynx *via* the oesophagus to the stomach, along which food is carried in mucus, aided by ciliary action. At this stage too the endocrine glands of the pharynx have appeared: the thyroid, which is paired, is already detached from the pharyngeal floor but the thymus is still a part of pouch 3 on each side.

During the second-form tadpole stage the pharyngeal roof becomes more complex, acquiring three dorsal longitudinal folds which increase its internal surface area. In the third-form tadpole the floor of the pharynx also becomes folded: it acquires ciliated bands which are thrown into folds at the level of each gill pouch. There are by now internal gills, and these bear epidermal ridges along their margins. These ridges or 'rakes' (Nieuwkoop & Faber, 1956) increase in number and acquire tooth-like projections which interlock between adjacent gill arches, thus helping to hold back food particles and direct them towards the pharynx. The movement of water and particles is maintained by continuous 'gulping' movements of the mouth and throat, which are controlled by powerful pharyngeal muscles.

In connection with absorption of food, the intestine becomes much elongated and coiled during the third-form tadpole stage. By the end of this stage, there are seven complete turns in the spiral intestine. This is in fact a double spiral, with the upper regions of the small intestine forming its outer coil and the distal regions leading into the colon, forming the inside of the coil. The stomach and pancreas are also well developed by this time and the internal structure of stomach and duodenum—gastric glands and typhlosole—have formed by stage 51, just before the hind limbs appear. The liver and gall bladder are functional a little earlier, at stage 47 (when all the yolk has been absorbed), and islets of secretory cells appear in the pancreas by stage 48. The midgut, interestingly, goes through a stage when its lumen is obliterated in anterior regions: this is due to excessive proliferation of epithelial cells, some of which slough off and block the lumen until they are removed by phagocytosis. (A similar process occurs in the development of the mammalian oesophagus and may occasionally lead to it remaining obstructed at birth.)

A general feature of the gut endoderm cells is that they retain yolk longer than other tissues of the larva (hence the yellow ventral colouring, noted earlier). By stage 47, however, all this yolk has been utilized and the larva is dependent entirely on its filter-feeding mechanism for its further nutritional needs.

(f) Urinary and Genital Systems

There are two successive kidneys in *Xenopus*, as in other amphibians: the transitory, small pronephros and the larger, definitive mesonephros. The pronephric kidney is already present at hatching, having developed from intermediate mesoderm lateral to the somites, in a few anterior segments of the abdomen. A pronephric duct leads straight back from each kidney to the cloaca. As the pharynx expands in the first-form larva, the pronephros is displaced caudalwards, however, so that its duct is no longer straight but exits from the medial wall of the pronephros and runs forwards, before turning back to continue caudalwards in the midline. The mesonephros appears in the second-form larval phase: its tubules are far more numerous than those of the pronephros and they open into the pronephric duct further caudally. By the third-form larval stage the mesonephros is elongated and functional, replacing the pronephros which now begins to degenerate. Strangely enough, besides the total degeneration of the pronephros at stage 53, there is also a degeneration of some of the mesonephric tissue at stage 55 (Nieuwkoop & Faber, 1956): this is an interesting phenomenon, not hitherto reported in vertebrates in which the mesonephros is the definitive kidney.

The gonads and gonoducts develop during larval life in *Xenopus*. As we saw in Chapter 4, the primordial germ cells migrate to the dorsal endoderm by stage 40 and reach the genital ridges by stage 43. Sexual differentiation is not apparent in the gonads till stage 52, however. The testis, in which the medullary layer develops, is a compact structure by stage 53–4. Seminiferous tubules are present by stage 55, and the first spermatocytes may be seen at stage 59. The ovaries, in which cortical tissue develops, are also compact structures by stage 53–4, and they contain oöcytes by stage 55. The oviducts (Mullerian ducts) do not appear until stage 64–66, however, when metamorphosis is nearing completion. Whether they develop at this same age (51–56 days after hatching) in neotenic larvae whose metamorphosis has been delayed, has not been reported.

Before going on to consider the changes that occur at metamorphosis, it is of interest to note the main features in the larva of *Xenopus* which Weisz (1945a) regarded as unique and unlike any other anuran larvae. These are: the large relative size of the head and its transparency: the mode of feeding and concomitant special structures in the pharynx for filtration and transport of food; the possession of lungs from hatching onwards; the dorsal extensions of the lungs to make contact with the middle ear (see page 24); the losses of certain aortic arches including the original carotid arch; the drainage of the gills into the jugular veins instead of into an efferent branchial system; other details of the blood system and of the cranial ganglia which we have already noted in Chapter 2; and the transverse orientation of the coils of the intestine, in contrast to other anurans where they lie longitudinally.

3. The Main Morphological Changes at Metamorphosis

Externally, the obvious signs of metamorphosis in any anuran larva are the appearance of limbs and the disappearance of the tail. These events can be seen very clearly in *Xenopus* too. Figure 5.13 gives sketches of the overall changes in body form that take place. The stage usually referred to as the 'metamorphic climax' is when the forelimbs emerge and the tail begins to shorten. In *Xenopus* this occurs at stage 59–62. These external changes are accompanied by changes in the head skeleton, the vascular system, the pharynx and the gut. We will deal first with the external changes, however.

(a) Resorption of the Tail

This begins with the degeneration of the distal tip of the notochord at stage 61. The degeneration process extends gradually to more proximal regions of the tail, reaching the proximal end by stage 62. Next, the connective tissue round the notochord, including cartilaginous rudiments of vertebrae, disappears gradually, followed by degeneration of outer tissues progressively from tail tip to base, and deposition of melanin under the skin. The muscles are the last to degenerate, but do so quickly, so that they have disappeared by stage 65, and the tail is represented only by a small dorsal swelling at stage 66. The ultrastructure of the degenerating muscles has been described by Weber (1964). This work, and other investigations on the biochemical events in the tail, will be referred to in Chapter 10.

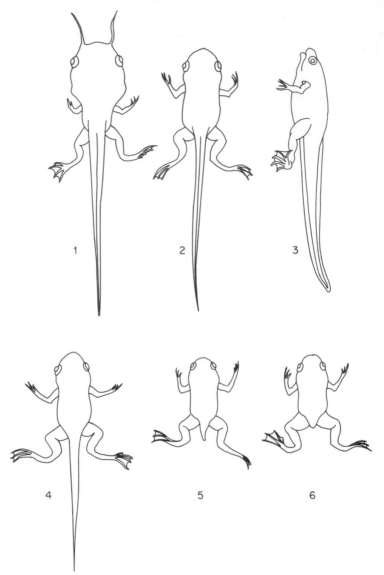

Figure 5.13. Diagrammatic sketches of the final stages of metamorphosis in *Xenopus*, when the tail is resorbed. From Rich and Weber, 1968

(b) The Development of the Limbs

The hind limb develops a little earlier than the forelimb, and is evident as a bud laterally by stage 46–47 (3 days after hatching). The pelvic girdle begins to develop at stage 50, and the limb is innervated by stage 51. By stage 66, when metamorphosis is finished, the pelvic girdle is complete and has ossified. Development of the hind limb carries on parallel with this, in proximo-distal sequence, ossifying in this sequence too. The femur begins to ossify at stage 55, the tibia and fibula at stage 56 and the foot skeleton from stage 56–66. The phalanges ossify before the tarsals, however.

The forelimbs of *Xenopus* develop in sacs which separate them from the gill chamber (Newth, 1949). This is unlike other anurans and is possibly a specialization to allow filter feeding to continue during the emergence of the forelimbs. Well-defined forelimb buds are present by stage 48, and are innervated by stage 51, when the rudiments of the shoulder girdle can be distinguished in mesenchyme. The shoulder girdle is complete by stage 60. In the forelimb itself, ossification has begun in the humerus and in the radio-ulna by stage 57, while the metatarsals and phalanges ossify by stage 58. It is clear, then, that the development of the forelimb is in the end more rapid than that of the hindlimb, and that by the time it emerges through the operculum of the gill chamber at stage 59, its ossification is already complete.

(c) Cranial and Pharyngeal Skeleton

Ossification in the skull, which occurs during metamorphosis, is accompanied by increasing fenestration both dorsally and ventrally, so that eventually only a narrow midline bony casing surrounds the brain, and this is connected to thin bony jaws laterally (cf. Figure 2.1 above). These changes are associated with great expansion in size of the special sense organs, which remain protected only by cartilage. There are also changes in the articulations of some of the bones at the postero-lateral angles of the skull with those of the jaws. The original larval lower jaw is formed by a cartilage similar to that in fish—Meckel's cartilage. This becomes displaced caudally and is replaced by membrane bones at the end of metamorphosis: the dentale and the goniale. The maxillary and premaxillary bones of the upper jaw have also ossified by this stage.

Cartilages of the gill region are much reduced in size at metamorphosis, but vestiges of them remain, as in other vertebrates, to form the hyoid apparatus at the base of the throat, and the laryngeal cartilages. In the male, the larynx enlarges later when sexual maturity is reached and it is able to croak, but at metamorphosis both sexes have larynges of similar size.

The vertebral column undergoes complete ossification and the transverse processes of the vertebrae are complete, by the end of metamorphosis. At the level of vertebrae 9 to 12, a rod of cartilage, the urostyle, develops during late larval life and this also ossifies, thus strengthening the vertebral column at hindlimb levels, before metamorphosis.

(d) Vascular System

Those changes that take place in the arterial system at metamorphosis have been depicted already in Figures 2.27 and 2.30. The ductus caroticus disappears, and aortic arch 3 narrows, whereas arch 4 increases in diameter and becomes the adult aortic arch. Other arteries which disappear are those of the filter apparatus in the gills. A new, cutaneous artery develops in connection with the pulmonary artery.

In the venous system, as we saw in Chapter 2, the mesonephric sinus takes over from the pronephric sinus which degenerates when the pronephros disappears. The anterior abdominal vein becomes prominent and leads into a large sinus at the ventro-caudal end of the abdominal cavity. Drainage later is from this region forwards, into vessels connected with the gut and liver. In the head region, the branches of the jugular veins that drained the branchial region disappear, while new veins develop in connection with the thyroid and thymus glands. In the tail, the caudal vein disappears as the tail regresses.

The lymphatic system has become progressively more extensive, and lymph hearts develop at metamorphosis. These, as we saw in Chapter 3, connect with the subscapular and the femoral veins. They are also in close connection with the dorsal lymph sac.

(e) Pharynx and Oral Cavity

In the mouth region, one of the most striking innovations at metamorphosis is the appearance of teeth. These form three rows on the upper jaw, and develop enamel, from stages 55 to 64.

In the pharynx, the gill clefts begin to close at stage 57. The first branchial cleft closes first, and the second and third have closed by stage 60, just after the forelimbs have emerged. At stage 63, the operculum closes. Degeneration of the branchial tissue internally occurs soon after this. The folds of the filter apparatus in the pharynx also disappear, giving a larger capacity to this region. In the floor of the pharynx, a primitive tongue rudiment develops.

Other derivatives of the pharynx, already mentioned, are the thyroid, parathyroid and thymus glands. These reach their definitive positions, having broken away from the pharyngeal endoderm, well before metamorphosis. All of them undergo enlargement and appear to function more actively as the metamorphic climax approaches (see also Chapter 10).

(f) The Gut

The most obvious morphological change in the gut at metamorphosis is a reduction of the coils of the intestine, from seven to two. The pancreas also alters its position and now lies in closer relationship to the stomach than to the liver.

Radical histological changes occur throughout the gut tract. The whole of its original epithelium undergoes histolysis and new, secretory cells take the place of the ones which functioned in the larva. In the stomach, too, the inner epithelium is extensively broken down and renewed, and in the pancreas there is extensive remodelling of secretory tissue. Most of these histological changes occur during stages 57 to 63, which cover the period of the metamorphic climax. The altered histology is connected with the change in diet and feeding habits. The adult *Xenopus* ingests solid food and will therefore require fewer ciliated cells in its gut lining, but a more powerful battery of enzymes than were needed by the filter-feeding larva which took in minute particles only. By the time metamorphosis is complete, at stage 66, the whole of the gut epithelium has in fact been renewed.

4. Concluding Remarks

This account of development in *Xenopus* has necessarily been brief, but further details of experimental work concerned with many of the processes that have been described, will be dealt with in succeeding chapters. As many illustrations of embryonic stages as possible have been included, but it is not easy to illustrate the detailed anatomy of the larva. Because of this, just the main timing of events in larval development has been given, and it is hoped that those who require more detail will refer either to the excellent descriptions of Weisz (1945a, b) or to Nieuwkoop and Faber (1956) for further references. Earlier literature is listed by Zwarenstein *et al.* (1946) also.

6

Observations on Cleavage and Blastula Stages

In this chapter we shall discuss some selected ultrastructural and biochemical observations on embryos of *Xenopus* at cleavage and blastula stages, and also some of the evidence relating to the differentiation of the cells.

The process of cleavage in *Xenopus* has not been described in the same morphological and histological detail as in certain urodeles (newts and salamanders) which became popular laboratory animals at the beginning of this century, when experimental embryology was in its infancy. Urodele embryos are usually preferred for descriptive work with the light microscope, because their cells are slightly larger and less yolky than those of anurans and are clearer to see in histological sections. In ultrastructural work, on the other hand, small cells have the advantage that more of them can be included within the limited area of an electron micrograph. *Xenopus* embryos have been used quite frequently in recent years for electron microscope work, though mostly at stages later than cleavage because they are then less fragile and can be fixed with less distortion. Three types of finding from ultrastructural studies on cleavage stages are worth mentioning, however, as they throw a little more light on mechanisms of early development in amphibians generally.

1. Ultrastructural Features of Cleavage Stages

(a) Furrow-formation

As we saw in the last chapter, studies of a number of amphibians including *Xenopus* suggest that the first event in the formation of a cleavage furrow is a contraction process in the outer layers of the egg cytoplasm. The mechanism of this contraction is not yet known. Electron microscope work has shown, however, that this outer cytoplasm contains various sizes of microfilaments, lying perpendicular to the line of the furrow (Selman & Perry, 1970). The filaments are finer in *Xenopus* than in *Triturus* and *Ambystoma* (a newt and a salamander), and there was at first some doubt as to whether they were similar structures in all these genera. Bluemink (1971a, b) is convinced, however, that both thin and thick filaments, found in different amphibian species, are contractile and play a role in furrow-formation. An experiment which would test this possibility would be to inhibit the contraction of the filaments and then to see if cleavage was blocked. Bluemink did in fact try this experiment, using an agent known as Cytochalasin B which blocks the contraction of microfilaments in a wide variety of animal cells (Estensen, Rosenberg & Sheridan, 1971). Bluemink found, however, that Cyto-chalasin B did not stop the initiation of cleavage furrows in *Xenopus* embryos. It did, however, cause the furrow to be incomplete and to open out later. The microfilaments were not visibly damaged by the treatment at first, though, so Bluemink concluded that the primary effect of the agent was on the cell surface, and that only secondarily, having caused reopening of the furrows, did it affect the contraction of the microfilaments. There is considerable doubt still, as to whether or not the microfilaments are involved in furrow-formation: a great deal more experimental work is needed. The nature of the material composing the various sizes of microfilaments in various amphibians also requires further investigation. Ultrastructural studies on these mechan-isms of cleavage are still being pursued (e.g. Bluemink & DeLaat, 1973).

(b) Origin of the Blastocoel

A further ultrastructural feature which is of general interest has been seen during cleavage in *Xenopus*, and not so far in any other species. Kalt (1971) observed that in *Xenopus* the formation of the blastocoel really begins at the first cleavage. He noted that small cavities remain at the tips of the furrows which separate the blastomeres on the animal pole side (Figures 6.1–6.3). These cavities gradually enlarge, partly because microvesicles at the surfaces of the cells bordering the cavities discharge their contents into them. Eventually several of these enlarged cavities at the bases of adjacent cleavage furrows in the animal half of the embryo become confluent and constitute the blastocoel. If Kalt's interpretations are correct, the blastocoel should contain considerable quantities of β-glycogen, since this is the main product dis-charged into it from secretory vesicles of the blastomeres. Whether this glycogen plays any important role later is a very interesting question. Early histochemical studies on other amphibian species (Woerdeman, 1933; Jaeger, 1945) showed large-scale breakdown of glycogen in the tissues invaginating at the beginning of gastrulation. Glycogen has also been seen in the interspace between dorsal mesoderm and ectoderm of the gastrula (Tarin, 1971b), which may represent remainders from the blastocoel. There has been some speculation as to whether it plays any role in neural induction (see Chapter 8).

Figures 6.1–6.3
Electron micrographs showing the contacts and spaces between cleavage cells. From Kalt, 1971.

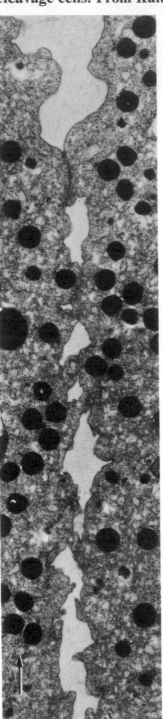

Figure 6.1. Junction zone between two cleavage cells, showing several gaps which become filled with fluid of the future blastocoel, and also some close junctions. Linear magnification × 7,600

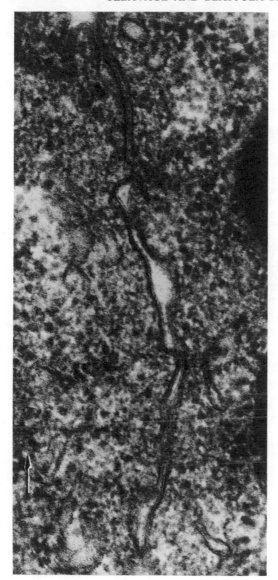

Figure 6.2. Detail of the close junctions. Linear
magnification × 105,000

(c) Ultrastructural Differences between Cleavage Cells

The electron microscope studies carried out so far on cleavage stages of
Xenopus have not added much to our knowledge about the earliest processes
of differentiation in the cells. Observations have been limited by the few

histochemical techniques available in electron microscopy, and the distributions of RNA and glycogen have occupied most attention. Van Gansen (1967) traced the location of granules containing RNA and glycogen, through 2-cell, 4-cell, morula and blastula stages of *Xenopus*. He, too, noted that glycogen granules accumulated between mesoderm and ectoderm at the

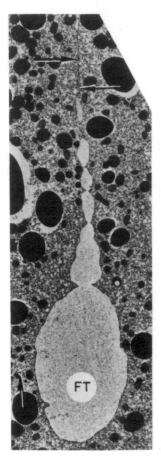

Figure 6.3. Dilatation (FT) at the base of a cleavage furrow. Note that it contains fine granular material, which may be glycogen. Linear magnification × 3,000

onset of gastrulation. Van Gansen saw cytoplasmic processes, containing granules which he regarded as polyribosomes, invaginating into the nuclei in some of the dorsal cells during cleavage. These did not appear in ventral cells, so they may indicate that some process of differentiation is occurring in dorsal cells at this stage. However, these observations have not been pursued further in cleavage stages.

2. Biochemical Observations at Cleavage Stages

Most of the biochemical work on *Xenopus* embryos has been applied to gastrulae and later stages of development, since these provide more material for analysis. In early cleavage stages, little net synthesis of materials occurs, and it would require very highly sensitive techniques to detect biochemical differences between different cells or regions of the embryo. It is a considerable achievement that Claycomb and Villee (1971) have been able to identify four subunits of the enzyme lactate dehydrogenase in eggs and early cleavage stages of *Xenopus*. Possibly differences in these subunits may be found in different blastomeres, when more sensitive methods are devised for detecting them. If so, this would be evidence for differences in gene activity in the different blastomeres, since according to current theory, each functional genetic unit or 'cistron' controls the production of a polypeptide subunit of a protein.

The statement that little synthesis occurs during cleavage stages in *Xenopus* applies also to DNA. This fact has puzzled investigators, since the nuclei are dividing during this period. Much of the biochemical work has focused on what extranuclear sources may provide DNA precursors for the dividing nuclei.

The DNA Balance Sheet

In all somatic cells (i.e. those of the animal's body which are not part of the germ cell line), the nucleus doubles its DNA content before it divides, at each mitosis. When the total DNA content of amphibian embryos is measured during cleavage, however, it is found that there is very little change: no overall increase in DNA is apparent until the onset of gastrulation. This is because during the several mitoses of cleavage nuclei, the DNA doubling is obscured by the presence of a relatively enormous reserve of DNA in the cytoplasm, which is only gradually depleted as the embryo develops. We do not know how much, if any, of this reserve DNA is taken up into the nuclei eventually. But the reserve has been estimated to be equivalent to several thousand diploid nuclei, as scrutiny of Table 6.1 shows. The first observations of this enormous DNA store in eggs were made on sea urchins by Hoff-Jørgensen and Zeuthen (1952) and on the frog *Rana pipiens* by Sze (1953). It has since been confirmed that the same situation holds in *Xenopus* (Vilimikova & Nedvidek, 1962: Bristow & Deuchar, 1964). Bristow and I found, however that there was a slight, steady rise in the total DNA content of the embryo during cleavage: the rise did not start only at gastrulation, as reported by other workers. So it looks unlikely that cytoplasmic DNA is the entire source of nuclear DNA during cleavage.

There has been much controversy as to the location of the cytoplasmic DNA in amphibian embryos. Biochemical and cytochemical work on

Table 6.1. Estimates of total DNA (μg) per cell in frog embryos.

Stage of development	*Rana pipiens* (Sze, 1953)	*Xenopus laevis* (Bristow & Deuchar, 1964)
Zygote	0·96	0·52
2-cell stage	0·51	–
4-cell stage	0·25	—
Morula	—	$0·15 \times 10^{-1}$
Blastula	$4·2 \times 10^{-4}$	$3·2 \times 10^{-4}$
Early gastrula	$4·1 \times 10^{-5}$	$2·6 \times 10^{-5}$
Late gastrula	$2·5 \times 10^{-5}$	—
Early neurula	$2·4 \times 10^{-5}$	$1·3 \times 10^{-5}$
Late neurula	$1·7 \times 10^{5}$	$0·88 \times 10^{-5}$
Tailbud stage	—	$0·84 \times 10^{-5}$
Hatching larva	$1·4 \times 10^{-5}$	$0·84 \times 10^{-5}$
Adult liver nucleus	$1·0 \times 10^{-5}$	—

Xenopus has now shown that about 70 % is bound to yolk platelets (Baltus & Brachet, 1962) and a further 10 % is in mitochondria (Dawid, 1966). The mitochondrial DNA is circular rather than linear in form, and its base composition differs from other embryonic DNA but resembles mitochondrial DNA found in other eukaryote cells (Dawid & Wolstenholme, 1967). This very peculiar DNA seems unlikely to be a source for nuclear DNA. Whether the yolk-bound DNA is a direct source of nuclear DNA or not, has not been decided: no conclusive labelling experiments have yet been done. But there is a substantial pool of free deoxyribonucleotides (Pestell, 1971) in *Xenopus* embryos which seem the more obvious immediate precursors for DNA. Work by Tencer (1961) showed that blocking of the uptake of labelled nucleotides from the cytoplasm into the nuclei, by the inhibitor 5-fluoro-deoxyuridine, will also block cleavage. So evidently some uptake of precursors normally occurs and is essential for normal cleavage. The yolk platelets are progressively broken down during development, so that they could contribute further nucleotide precursors if their DNA is also digested in the breakdown process.

It is difficult to draw precise conclusions, or to make out an accurate quantitative 'balance sheet' for DNA in early amphibian development, because estimates obtained by different authors using different techniques to extract and measure it, have varied extremely widely: Table 6.2 gives a comparative survey of the data. It can be seen that these differ enormously between the different authors. Probably Dawid's data are the most reliable, since he obtained purified DNA preparations by separation on a caesium chloride gradient, whereas other authors' DNA extracts may have been contaminated with a variety of materials.

What these data do show consistently, despite their differences, is that the

Table 6.2. Estimates of the total DNA (μg/embryo) in *Xenopus laevis*

Stage of development	Vilimikova & Nedvídek, 1962	Bristow & Deuchar, 1964	Dawid, 1965
Egg	0·46	0·52	—
2-cell stage	—	—	0·01
4-cell stage	0·58	—	—
Morula	0·60	0·75	—
Blastula	0·82	0·79	0·09
Early gastrula	0·99	1·00	0·20
Mid-gastrula	1·27	—	—
Late gastrula	1·45	—	—
Early neurula	—	1·30	—
Mid-neurula	1·77	—	0·30
Late neurula	—	1·60	—
Tailbud stage	—	1·80	0·40
Hatching larva	—	2·10	—

total DNA *per cell* decreases during early development—i.e. the cytoplasmic reserves evidently get used up—until by about the tailbud stage, just before hatching, the *per cell* value has fallen to that normal for an adult diploid nucleus (cf. Table 6.1). Now although Lohmann (1972) has recently shown some short-term variations in nuclear DNA content in different tissues of the newt *Triturus vulgaris* from gastrula stages onwards, it is generally assumed that throughout cleavage, embryonic nuclei all contain equal quantities of DNA. There are certain organisms, for instance some of the insects, in which some embryonic cells lose chromosomes during cleavage and therefore do change their nuclear DNA content: but in by far the majority of animals studied, no such radical changes occur. It remains possible, however, that *qualitative* nuclear changes take place during cleavage, which lead to differentiation of the cells. This is a point which we must now consider in more detail.

3. Evidence on the Extent to which Cleavage Cells are Differentiated

One of the most intriguing questions for embryologists is how far the individual cells of the cleaving embryo are already different in their developmental potentialities. In many invertebrates—molluscs and annelid worms, for instance—each cell has very clearly defined destiny from the moment it has formed, and removal of that cell will result in absence of a particular organ or tissue from the larva. As a corollary to this, individual cells isolated after the first few cleavages have occurred will form only parts of embryos: there is no possibility of one cell adapting and reconstituting the whole organism. In other invertebrates, however (e.g. echinoderms) and in most vertebrates, one isolated blastomere even as late as the 8- or 16-cell stage may be able to form a whole organism. This shows that in these animals there are no changes in the nuclei at early cleavage stages, such as to limit permanently the potentialities of the cells for differentiation. This is such a fundamental point that it is worth digressing first to experiments on other animals, before discussing the recent and important findings in *Xenopus laevis*.

The classic experiment which showed the equivalent potentialities of the first two blastomeres in amphibians was carried out by Spemann at the turn of the century. (An earlier attempt by Roux (1888) to follow the fate of one frog blastomere after killing the other by pricking, had given misleading results because the presence of the dead cell impeded the progress of the other cell). Spemann (1901) achieved a separation of the two blastomeres without killing either, by tying a ligature of hair round the two-cell stage embryo, in the newt *Triturus*. He found that both cells could form complete larvae which were just as normal as one developed from an unoperated zygote: they were no smaller than normal, either. There were similar findings (Driesch, 1910) in sea urchins, whose cells may be separated by placing the embryos in calcium-free sea water—and also in Nemertine worms. The blastomere separation experiment was achieved in mammals too: Seidel (1960) and Nicholas and Hall (1942) were able to obtain single blastomeres from two-cell stage rabbit embryos which had been washed out of the uterus. Seidel's procedure was to kill one cell by cautery with a hot needle, whereas Nicholas and Hall managed to separate the two cells by up-and-down suction in a narrow pipette. In each case the live cells were replaced into the uterus of a female rabbit with genetically different coat colour from that of the strain from which the cells came. So the presence of a differently coloured individual in her litter showed that the foreign blastomere had developed, implanted and been carried to term, producing a complete and normal individual. More recently Tarkowski and Wroblewska (1967) have carried out similar blastomere separation experiments on mouse embryos. These also carried genes for coat-colours differing from those of the mother into which the cells were then introduced. A low proportion of viable embryos were obtained from blastomeres separated at the 4- and 8-cell stage as well as at the 2-cell stage. So it can be concluded that in mammals, as in the majority of the other animal embryos tested, every one of the early cleavage cells is potentially capable of forming a whole embryo. There is no evidence of any irreversible differentiation having taken place in the cells.

Since differentiation in a cell originates from events taking place in its nucleus—i.e. from gene activity—a more direct approach to the investigation of how far the cleavage cells are differentiated is to test the potentialities of their nuclei. If a blastomere nucleus has not differentiated in any way from the condition of the original zygote nucleus, it should be able, when transplanted into the cytoplasm of an egg, to initiate normal development and form a complete embryo. This possibility has been tested in amphibians, by the now famous and classic *nuclear transplantation* experiments. The earliest of these experiments were carried out on *Rana* by Briggs and King (1952) and by Moore (1960), and since then much more extensive nuclear transplantation work has been done on *Xenopus*, initiated by Elsdale, Gurdon and Fischberg (1960) and carried on by Gurdon and his collaborators at Oxford. The principles of the experiments are shown in Figure 6.4. This

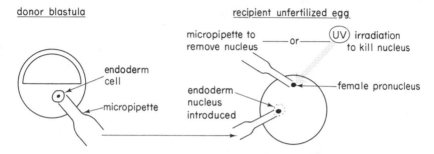

Figure 6.4. Principles of the nuclear transfer experiment

work has been very well reviewed by Gurdon (1964), so we shall just state the main findings here. These were, that in both *Rana* and *Xenopus*, the nucleus from an endoderm cell of the blastula, injected into the egg after removal of the egg nucleus, will produce completely normal embryonic development in a good proportion of cases (25–30 %). Ectoderm and mesoderm nuclei will also give rise to normal embryos, but a lower percentage of these transplants are successful. This may be because there is less cytoplasm in these smaller cells to protect the nucleus from damage in the micropipette used for the transfer operation. Another general finding was that nuclei taken from progressively older embryos gave progressively lower percentages of normal embryos. This indicates that *some* irreversible changes do accumulate, in at least *some* of the nuclei, as development proceeds.

It has also been found possible to discriminate differences between the individual nuclei of a single embryo, by raising clones of embryos derived from a series of nuclear transplants through several generations of descendants from each nucleus. This 'serial transfer' experiment was devised by King and Briggs (1956) and is illustrated in Figure 6.5. Embryos or larvae of different clones, derived from the different endoderm nuclei of a single original blastula, show characteristic differences in morphology. This implies that there were stable genetic differences between the original endoderm nuclei.

Some wider applications of the nuclear transplantation method are worth mentioning here. It provides a way of testing interactions between nucleus and cytoplasm, as well as testing the stability of genetic differences. Gurdon (1960a) used serial transfer experiments to test the stability of differences

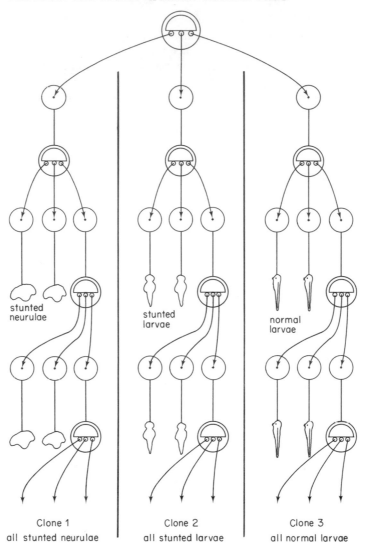

stunted
neurulae

stunted
larvae

normal
larvae

Clone 1	Clone 2	Clone 3
all stunted neurulae	all stunted larvae	all normal larvae

Figure 6.5. Principles of the serial transfer experiment: modified from King and Briggs (1956)

between embryonic nuclei of different subspecies of *Xenopus laevis*. Nuclei of *Xenopus l. laevis* were transplanted into the egg cytoplasm of *X. l. victorianus*, for several generations, and it was found that the developing larvae and frogs persisted in showing only the characters of the nuclear subspecies, and not any features attributable to factors from the cytoplasm. So evidently the differences between nuclei of these two subspecies are very stable. Nuclear transplants have also been tried between *Xenopus* and other genera of anurans: for instance between *Xenopus* and *Rana* (Fischberg, Gurdon & Elsdale, 1958), between *Xenopus* and *Hymenochirus* (Gurdon, 1959) and between *Xenopus* and *Discoglossus* (Woodland & Gurdon, 1969). In none of these cases did development go beyond a few cleavages. But it could at least be shown that the cytoplasm of amphibian eggs promotes replication of nuclei from foreign cells. Gurdon, Birnstiel and Speight (1969) went on to show that even raw DNA, when introduced into *Xenopus* eggs, was able to replicate.

In their later work, Gurdon and his collaborators made nuclear transplants into *Xenopus* eggs from quite advanced tissues such as the intestinal epithelium of the larva, or from adult skin, liver and other organs which had been grown in tissue culture and remained in the same differentiated state for long periods. Very few, but at least *some* normal frogs were obtained from these transplants (Gurdon & Laskey, 1970). So it has to be concluded that in certain circumstances it is possible to reverse even those nuclear changes which are normally stable indefinitely in tissue culture. The reason why some nuclei give normal embryos and others do not, has not yet been traced to initial differences in the nuclei, but other workers have sometimes observed damaged chromosomes and abnormal mitoses after nuclear transplantation, which might be caused by the transplant procedure. It is of course very difficult not to damage the nuclei in these very delicate micromanipulations.

One reason for the ability of nuclear changes to be reversed, and of genes to 'switch on' their early developmental functions again when an older nucleus is put into an egg, may be that the egg cytoplasm contains some special stimulating factors. Gurdon (1968) showed that brain nuclei which had normally ceased both DNA and RNA synthesis would enlarge and start synthesis again when placed in egg cytoplasm. Even more surprisingly, they lost their nucleoli, like the egg nucleus, and did not acquire more nucleoli unless or until the embryo developed to a blastula stage (when nucleoli normally reappear, cf. Chapter 4). But although the egg cytoplasm does therefore seem to have major effects on nuclear function, we saw from Gurdon's transfer between subspecies that there is no evidence that genetic characters are altered at all: the cytoplasm is simply a stimulator of the normal genetic activity in the transplanted nucleus. In fact we now have evidence that the egg cytoplasm is ready to 'obey the instructions' of a foreign nucleus and to synthesize quite foreign proteins. Gurdon, Lane, Woodland and Marbaix (1971) have shown that if the mRNA for rabbit haemoglobin is injected into a *Xenopus* egg, this will synthesize rabbit haemoglobin.

For the embryologist interested in cell differentiation at early stages of development, the most important outcome of all the nuclear transplantation work is that it shows that, given favourable circumstances, the genes of early embryonic cells can repeat all the activities they have already carried out, from the zygote stage onwards and, in egg cytoplasm, are potentially able to form a complete new embryo. What the small genetic variations are which make some transplanted nuclei more successful than others, we still do not know.

4. Control Processes during Cleavage

(a) The Grey Crescent

We have seen already (Chapter 4) that in amphibian embryos the future dorsal side is determined at fertilization. A 'grey crescent' of lighter pigmentation than the rest of the animal hemisphere appears, usually diametrically opposite to the point of entry of the sperm. The crescent forms even in unfertilized, parthenogenetically activated eggs, however, so its formation is evidently governed by maternal factors and not by any essential factors in the sperm. It has been shown in *Xenopus* by Curtis (1960a, 1962) that the grey crescent controls the cleavage pattern and subsequent orientation of the embryo by some process which is complete before the 8-cell stage. Curtis succeeded in grafting pieces of grey crescent cortex into the ventral regions of other embryos and found that when this was done prior to their third cleavage, they were induced to form a secondary axis ventrally. But if the graft was made at the 8-cell stage or later, it had no effect. The role of the grey crescent in controlling axis-formation has been investigated further recently by Grant and Wacaster (1972), who have shown that the controlling powers of the crescent are destroyed by irradiating this region with ultraviolet light. After such treatment to an embryo at cleavage or blastula stage, no neural axis formed at all. These authors went on to show that nuclei which had been transplanted from the irradiated embryos into normal egg cytoplasm, were perfectly capable of initiating normal development: so the grey crescent's effect which had been destroyed, had nothing to do with these nuclei. On the other hand, it was possible to restore irradiated embryos to normal and induce an axis to form, by injecting normal grey crescent cytoplasm into the blastocoel at the blastula stage. This is a remarkable finding and indicates that the grey crescent cytoplasm may contain some 'inductor' substance which controls later processes as well as the initial orientation of the embryo during its first cleavages. Presumably this later influence is that normally exerted by the cells of the dorsal mesoderm, into which this grey crescent cytoplasm becomes incorporated. This topic of neural induction by the mesoderm will be discussed in Chapter 8.

(b) Communication between Cells

As cleavage progresses, it becomes impossible for any one region of cytoplasm, like the grey crescent, to have direct continuity with all parts of the embryo. This does not mean, however, that there can be no interchange of materials between cells. As we saw earlier, it is some time before the cell membrane between any pair of daughter cells is complete and there is at first a wide zone where their two cytoplasms are confluent, between the tips of the cleavage furrows as one views these in transverse section (cf. Figures 6.1–6.3). But even when the membrane is completed, there are junctions, visible with the electron microscope, which could allow the passage of small molecules between adjacent cells. Some of these close junctions are shown in Figure 6.2. It is possible that other kinds of communication, such as electrical signals analagous to nerve impulses, may pass across junctions of this kind, from cell to cell. Furshpan and Potter (1968) first drew attention to the existence of electrical discharges between the cells of a wide variety of animals and their embryos, and they believed that these discharges passed across at 'septate' junctions, which appeared to have pores through which molecules of up to 69,000 molecular weight might be able to pass. Electrical discharges have also been detected passing between the cells of blastulae of *Xenopus*, by Palmer and Slack (1969), and these are thought to represent some kind of intercellular communication which is essential for normal development. If the cells are electrically uncoupled, by treating the embryo with an uncoupling agent known as halothane, development is arrested.

The cells of the late blastula also intercommunicate, and at the same time express some degree of differentiation, by cell movements and behavioural interactions. These are readily demonstrated in cultures of the cells, after they have first been dissociated by treating them with a chelating agent such as 'versene' (ethylene-diamine-tetraacetic acid) which binds calcium and magnesium ions. When returned to normal saline, the cells of each layer—(ectoderm, mesoderm and endoderm)—show characteristic differences in behaviour. Ectoderm cells aggregate the most rapidly, and form epithelia which then spread outwards. Mesoderm cells show active pseudopodial activity, more marked than in ectoderm cells, but do not adhere quite so rapidly. They are soon able to undergo morphogenesis, some elongating to form a notochord and others forming the rudiments of somites. If in large enough numbers, they may also 'induce' some of their number to form an ectoderm layer round them and even a neural plate subsequently. Endoderm cells, on the other hand, are passive in culture, remain spherical and only slowly reaggregate into loose groups and epithelia.

Besides these differences in individual behaviour, the cells of different layers of the late blastula show different adhesive affinities, ectoderm sticking preferentially to ectoderm, mesoderm to mesoderm and endoderm to endoderm. These differences enable them to 'sort out' into their own tissues, after being mixed randomly. Most of the work on this sorting out process has been done on cells of the gastrula stage, and will be discussed in the next chapter, since it is relevant to the mechanism of gastrulation. In fact, the so-called 'blastulae' used for demonstrating differences between ectoderm, mesoderm and endoderm at this supposedly earlier stage, may already have had some dorsal mesoderm below the surface, in *Xenopus* (cf. Chapter 5, page 108 ff.), so ought perhaps, strictly speaking, to be classed as early gastrulae.

These processes of intercommunication between regions of the cleaving embryo, first from the grey crescent and then as individual interactions between cells, provide the mechanisms by which embryonic development remains a coherent process. Work on cleavage and blastula stages in *Xenopus* has shown us a little more of how the mechanisms of cleavage, the gene activities and the cytoplasmic communication systems are so balanced as to produce a normal embryo, even when experimenters interfere at various points in the chain of developmental processes.

In the next two chapters we shall go on to see more of the mechanisms of gastrulation and neurulation in *Xenopus*, and of the interactions that occur between the tissues of these stages.

7

The Mechanisms of Gastrulation and Neurulation

As we have already seen in Chapter 5, gastrulation and neurulation are the phases of development characterized by the most extensive cell and tissue movements. Gastrulation, particularly, is marked by major changes in the relative positions of the embryonic cells. This is true of development in all vertebrates, and to some extent in invertebrates too, although they tend to have fewer cells and therefore to show smaller-scale movements than those of the vertebrates.

The controlling mechanisms which govern these early movements of embryonic cells have long been of interest to embryologists, who have investigated them by techniques ranging from detailed histology of whole embryos, to physical and biochemical studies on isolated cells. We owe much to the ingenuity of those who have devised new methods for handling and making measurements on these very small objects—individual embryos or even single cells or nuclei. In fact, experimental embryology would never have made the progress it has, if it had not been for the development of techniques for microsurgery and tissue culture, invented by Ross Harrison in the early years of this century. Since then, biochemists, physicists and engineers have added their expertise, till it is now almost necessary for any intending embryologist to have a smattering of training in all these fields. Amphibian embryos are still the most popular material for investigations on gastrulation and neurulation, because their cells are larger and the cell movements are more clearly seen than in other vertebrate embryos. Earlier work was done on amphibian species other than *Xenopus*, but a substantial amount of recent work has used *Xenopus* as the main experimental material. It will be remembered from Chapter 5 that gastrulation and neurulation differ in certain points of detail from these processes in other anurans that have been described. But the differences are not great enough for it to be thought that the underlying mechanisms are very different, and one may justifiably generalize from findings in several different species.

1. The Control of Gastrulation and Neurulation Movements in *Xenopus*

(a) Observations on Intact Embryos

A certain amount can be inferred about mechanisms of cell movement, from descriptions of the static appearance of cells in histological sections of whole embryos taken at a series of stages in gastrulation or neurulation. It is on studies such as these, of course, that the descriptive work dealt with in Chapter 5 is based. The advent of electron microscopy, however, has made it possible also to look in far more detail at changes in the surface structure of the cells and at any relevant changes in their internal contents. In addition, autoradiographic methods enable one to label certain cells, more accurately than used to be possible with various dyes, and hence to follow their movements and fates in some detail. Thus, Sirlin (1956) has been able to confirm Nieuwkoop and Florschütz's observations (1950) that the dorsal mesoderm is already below the surface in *Xenopus* embryos before the dorsal lip is visible. Sirlin allowed radioactively labelled metabolites to be taken up by cells lining the dorsal lip, and found that the labelling appeared later in dorsal *endoderm* only, and not in mesoderm. Tarin (1971b) has again confirmed this, from histological and electron microscopical studies of events during gastrulation. He noted that the cells which invaginated into the dorsal lip mostly lay posterior to the mesoderm, and formed endoderm, whereas cells already lying deep to the dorsal lip when this first formed gave rise to dorsal mesoderm (cf. Figure 5.6 above). Since no observer has yet seen any evidence that these mesoderm cells have invaginated from the surface, it appears that they must become a separate layer from the ectoderm simply as a result of mitoses in a tangential plane, and not by cell movements at this stage. They do, however, spread forwards under the ectoderm during gastrulation, just as in other amphibians.

One of the major puzzles in embryology is what *initiates* gastrulation. Earlier workers thought that it might be due to some osmotic effect, since embryos placed in hypertonic saline could not gastrulate, but underwent instead a reverse process of 'exogastrulation' in which all the mesoderm and endoderm turned outwards (Holtfreter, 1933). During gastrulation in all amphibians, a new internal cavity forms, the archenteron or future gut cavity. As the dorsal lip deepens, it creates a deepening groove which becomes this new internal cavity (cf. Figure 5.6 again). At the same time as the archenteron forms, the original cavity of the blastula (blastocoel) becomes gradually obliterated. It is still not certain where the fluid contents of the blastocoel go to. Tuft (1961, 1965) has, however made a physical study of the movements of water during gastrulation in *Xenopus*. Using a formula which relates 'reduced weight' (i.e. weight in water) to volume and density, he has been able to calculate the directions of net flow of water in the embryonic cells. He found that water evidently flows from the blastocoel into the archenteron, passing through the archenteron roof mesoderm and endoderm to do this. At neurulation, he calculated that some of this water passes into the neural tube, via the neurenteric canal, at the time when the brain vesicles expand. Tuft's diagrams are given in Figure 7.1. Unfortunately

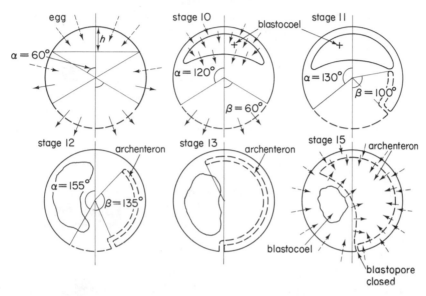

Figure 7.1. Diagrams showing the directions of flow of water in *Xenopus* embryos, as seen in sagittal sections. After Tuft, 1965. Solid arrows indicate energy-coupled water flow; dotted arrows, flow by osmosis. Note that at stage 10 (early gastrula) water enters the blastocoel, and that at stage 15 (early neurula) water enters the archenteron

these findings do not so far give a clear lead as to what may *initiate* the movements either of fluid or of cells. Prior to gastrulation, there seems to be a continuous inflow of water at the animal pole region and outflow at the vegetal pole (Figure 7.1), but Tuft did not note any alterations in flow at the region of the future dorsal lip. Other workers have attempted to see if disrupting the water balance by altering the permeability of the tissues will alter the course of gastrulation. Haglund and Løvtrup (1965) used digitonin for this purpose, but unfortunately found that the dosages which were effective were also lethal to the embryos. Another agent which has been found to upset the fluid distribution in *Xenopus* embryos is β-mercaptoethanol. Malpoix, Quertier and Brachet (1963) found that embryos treated with this agent suffered a collapse of the archenteron cavity and an accumulation of fluid ventrally between ventral ectoderm amd mesoderm. In these embryos, gastrulation was arrested. So evidently the normal movements of fluid are an important part of the gastrulation process. Interestingly, Hamilton and Tuft (1972) have found that in haploid *Xenopus* which later often become oedematous owing to inefficient functioning of the kidneys and inadequate water clearance in larval stages, also fluctuate in their water content during gastrulation. They may show losses of fluid through the blastopore. Ectoderm explants also absorb water more rapidly than in diploids. These differences are thought to be due to the doubling in number of cells that occurs in haploid gastrulae, to compensate for the smaller cell size.

The process of neurulation in *Xenopus* has also come in for some study, both descriptive and experimental, in recent years. Among the general theories applied to explain neurulation in amphibians, it has been suggested that there is a differential accumulation of water at the ventral ends of the neural plate cells (see Glaser, 1914), a contraction of their upper, dorsal ends (Balinsky, 1960) and an increase in their side-to-side affinities to different extents across the width of the neural plate (Jacobson, 1968). As we shall see, the recent evidence indicates that all three of these processes may be involved, in *Xenopus*.

The most detailed recent descriptive studies on neurulation in *Xenopus* are those of Schroeder (1970, 1973). He has shown by light and electron microscopy that both upper and lower layers of the neural plate are involved in the rolling up and formation of the neural tube. This is in contrast to reports by Nieuwkoop and Florschütz (1950) and Tarin (1971) that the upper layer forms epidermis. Ave and co-workers (1968) reported that the upper layer is subject to selective cell death after treatment with neural inductors (cf. Chapter 8). Schroeder's conclusions seem based on careful observations, however and generally acceptable. He concludes, as illustrated in Figure 7.2, that the main cause of neurulation is a contraction of midline cells of the neural plate in a transverse plane. In addition, the neural plate cells elongate in a dorsoventral direction, and the somite cells beneath them do the same. The mid-ventral surface of the neural plate becomes firmly attached to the notochord below it, and its later elongation is aided by extension of the notochord cells at the end of neurulation.

The major component of the neurulation movements, according to Schroeder's account, is the contraction of the midline cells. There have been many attempts to explain this in terms of known contractile mechanisms in other animal cells. The possibility that a myosin-like protein was present was tested on other amphibian species by treating the neurulae with ATP (Ambellan & Webster, 1962). This nucleotide appeared to speed up neurulation, so it was suggested that it enhanced the activity of a myosin-bound ATP-ase in the neural plate cells. However, there is considerable doubt as to whether ATP could enter the cells, since it is a relatively large molecule. So its effects may have been quite unconnected with any myosin-like protein, and may simply have involved some energy transfer to the cell surfaces. Other fibrillar proteins have been thought to be involved in neurulation too: Jacobson (1970) showed that in *Ambystoma* (the axolotl) neurulation was inhibited by β-mercaptoethanol, and this time it was argued that the agent was acting on disulphide bonds (—SS—), preventing the polymerization of fibrillar proteins. But as with the myosin theory, no one got as far as demonstrating the presence of these proteins in the neural plate, either by histochemical or by biochemical methods.

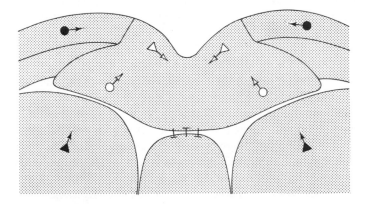

Figure 7.2. Diagrammatic transverse section of the early neurula of *Xenopus*, with arrows showing main directions of movements involved in neurulation. After Schroeder, 1970. White arrows: movements of neural plate; circled black arrows: converging movements in epidermis; triangled black arrows: movements in somites

Fibrils of various orders of size have been seen in neural plate cells of *Xenopus*, but so far their composition has not been established. Green, Goldberg, Schwartz and Brown (1968) identified collagen in extracts of neurulae of *Xenopus*, and were able by labelling with tritiated proline and by measurements of protocollagen synthetase activity to establish that collagen synthesis begins at the gastrula stage and is very active during neurulation. If so, it should have been possible for the electron microscopists to have identified collagen fibrils in cells of the neural plate if they were localized here. Transversely-orientated microfilaments have, indeed, been seen at the upper ends of the neural plate cells in *Xenopus* (Schroeder 1970; Baker & Schroeder, 1967; Balinsky 1960), but these have not yet been identified biochemically. In addition, microtubules, orientated vertically, have been seen in the neural plate cells (Schroeder, *loc. cit.*; Karfunkel, 1971), and it is thought that the vertical elongation of these microtubules could account for the change in shape of the neural plate cells from cuboidal to columnar. In support of this suggestion. Karfunkel showed that treatment with vinblastine sulphate destroyed the microtubules and also prevented the elongation of the cells, which resulted in a blockage of neurulation.

It can be seen from the foregoing account that the work on intact embryos has produced some descriptive, but rather less experimental evidence to explain the mechanism of neurulation. On gastrulation, not enough evidence can be gleaned about mechanisms from descriptive work, but rather more experimentation has been carried out on this stage of development. The most critical experiments are those that have been carried out on gastrula cells isolated from the embryo, however. We shall now deal with these studies in more detail.

(b) Observations on Isolated Groups of Cells

One of the most remarkable features about cells of the amphibian gastrula is that even in isolation from the rest of the embryo, they will show movements and behaviour that are relevant to their normal roles in gastrulation. With gastrula cells (but not so easily with neurula cells, because these are more cohesive), it is possible to disaggregate them by placing pieces of embryonic tissue in saline which lacks calcium and magnesium ions. These two ions are essential for the adhesion of embryonic cells. They may be removed from the medium and from the cell surfaces by adding a 'chelating' agent such as Versene (ethylene-diamine-tetraacetate) which binds divalent ions. The factors controlling adhesion and movements of gastrula cells may then be studied in detail, having returned the loose cells to normal saline, in which they are able to stick together again. In *Xenopus* much of the initial work on reaggregation of gastrula cells was carried out by Curtis. He found (Curtis, 1957) that blastula and early gastrula cells take up more calcium on reaggregation than do cells of other stages, and that their rate of adhesion depends on the concentration of calcium in the medium. The calcium does not penetrate the cells, but is adsorbed onto their surfaces.

A striking property of gastrula cells is that they show differential affinities as they reaggregate, and they adhere preferentially to cells of their own germ layer. This, together with a motility which enables them to move about and 'find' cells of their own type, causes the three layers, ectoderm, mesoderm and endoderm to 'sort out' after being randomly mixed up in culture. Townes and Holtfreter (1955) were the first to demonstrate this, in cells of *Ambystoma* embryos, and they found that, out of the random mixture, ectoderm would come to lie on the outside, mesoderm in the middle and endoderm innermost, just as in a normal embryo. They explained this by a tendency of ectoderm cells to spread round the other cells, while mesoderm tended to invaginate into groups of cells of other types, and endoderm was somewhat passive and immobile. It was Curtis (1961) who suggested that the 'sorting out' of the three germ layers was also explicable on the basis of differences in timing of their reaggregation movements. Using mixtures of cells from *Xenopus* gastrulae, he showed that it was possible to make endoderm arrive on the outside, if it was disaggregated and returned to normal saline 6 hrs in advance of the mesoderm and ectoderm (Figure 7.3). It was then able to move out of the mixture before the other cells had become mobile, and to form an epithelium round them by virtue of its preferential adhesion to cells of its own type. So it seems likely that in the normal way when all cells are disaggregated simultaneously, the ectoderm emerges on the outside because it has become mobile more quickly than cells of the other two layers and has then adhered to other ectoderm cells on the outside of the mixture.

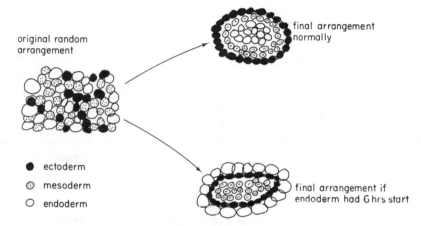

Figure 7.3. Diagram to illustrate the results of Curtis's timed aggregation experiments on blastula cells. (a), on left, original random mixture of cells. (b), top right, their final layout after reaggregating normally. (c), bottom right, endoderm comes outside if it has started reaggregation 6 hrs. earlier than the other cells

Besides timing mechanisms, specific chemical recognition mechanisms have been suggested as the basis for the movements and differential adhesions among gastrula cells. Much interest was at one time focused on a slimy material that exudes when cells are disaggregated, and appears to occupy the intercellular spaces. This intercellular material, or 'ICM', is now thought to consist mainly of mucopolysaccharides. But earlier work (Curtis, 1958) indicated that it could contain up to 25 % of RNA: so the possibility remains that the ICM could carry some kind of molecular information which enables the cells to recognize others of like type. A more popular suggestion has been that the cells carry surface marker molecules similar to antigens. There is evidence of antigenic differences between ectoderm, mesoderm and endoderm in early amphibian embryos, from tests with antisera on protein extracts of the cells (Clayton, 1953; Inoue, 1961) and from work with fluorescent antibodies applied to sections of tissue (Takayanagi, 1958; Stanisstreet, 1972).

But it is not possible so far to distinguish whether the tissue-specific antigens are carried on the surfaces of the gastrula cells or internally.

Recent electron microscope work has led to observations of various ultra-structural channels of communication between gastrula cells. Close junctions between the cell membranes occur quite frequently (cf. Figure 6.2 above) and the pores that have been seen in some of these are of sufficient diameter to allow molecules of up to 69,000 molecular weight to pass through them. Slack and Palmer (1969) showed, however, that this passage must be selective, for fluorescein, with a molecular weight of only 332, was not able to pass from cell to cell in *Xenopus* embryos. These same workers have shown that electrical impulses travel between the cells in the blastula, and it is thought that these impulses go through the close intercellular junctions. (Palmer & Slack, 1969: cf. Chapter 6). It is possible that such electrical signals could provide the initial stimuli for gastrulation and/or neurulation movements, or that they co-ordinate these movements once they have started. A great deal more experimental work is needed, however, before we can do more than speculate on the initial causes of gastrulation and neurulation. Cohen and Morrill (1969) have postulated that gastrulation in *Rana* is initiated by a potential difference set up between the blastocoel and the external medium, as a result of the passage of sodium ions inwards. They have shown that there is an inflow of these ions, but they have no evidence as to the cause of the inflow—so in fact the *initiating* circumstances are still left unexplained by their theory.

Another idea which has given rise to considerable testing and controversy is that the movements of cells in gastrulation bear some relation to their surface charge. Curtis (1960b) discussed the physico-chemical effects of surface charges on cell motility and adhesiveness. A high surface charge may result in repulsion between cells of like charge, but a low surface charge makes the cell membrane more readily deformed, and hence may aid movements that depend on continuous changes of shape. MacMurdo and Zalik (1970) found that among the cells of *Xenopus* gastrulae, mesoderm had a lower surface charge than either ectoderm or endoderm. This would perhaps provide an explanation for the ability of mesoderm cells to move rapidly and to invaginate in certain areas. In species where mesoderm is seen to invaginate at the dorsal lip, there are spectacular changes in cell shape here. These have been described at the ultrastructural level in gastrulae of the tree-frog, *Hyla regilla*, by Baker (1965). It is doubtful, however, whether any such invagination of dorsal mesoderm takes place in *Xenopus*, and as we noted in Chapter 5, there is little change of shape detectable in invaginating cells as viewed under the light microscope (cf. Figure 5.7 (b)).

It is still difficult to draw any conclusions from the work on isolated cells that can be applied at all confidently to the situations in whole embryos. Amphibians other than *Xenopus* are more favourable material both for the whole-embryo work and the cell cultures, simply because their cells are somewhat bigger. One line of approach that has also been possible in other amphibians is to study gene-controlled differences in cell movement. These hold important possibilities for the future, so we will digress to consider them briefly here.

(c) Studies on Hybrid Gastrulae

When trying to assess how far the results of cell culture work are relevant to events in the whole embryo, a possible approach is to try to correlate abnormalities of gastrulation with variations in the reaggregation behaviour of the cells. Interspecific hybrids, made by fertilizing an egg of one species with sperm from another, are sometimes unable to gastrulate and hence do not survive. Johnson (1969) made a study of reaggregation behaviour in gastrula cells of various hybrids of *Rana* species. He found, in general, that the more normal were the properties of aggregation of the hybrid gastrula cells, the more likely was it that the intact gastrulae would undergo normal gastrulation. Some hybrids were arrested at early gastrula stages, and in these the cells did not reaggregate normally; others were capable of developing beyond the gastrula stage, and the cells of these showed normal reaggregation. So there was an evident correlation between normality of cell behaviour and normality of morphogenetic movements in the whole embryo. This makes it more possible to believe that work on isolated cells throws useful light on normal morphogenetic processes. Another point that is highlighted by Johnson's work is that the behaviour of the gastrula cells is under direct genetic control, in the first instance. We shall see in the next chapter examples of ways in which the initial controls set in train by gene action are modified later by interactions between tissues that have been brought into contact by the movements of gastrulation and neurulation.

———————————

To conclude, though no single finding provides an obvious single cause of either gastrulation or neurulation, all of the intracellular structures and the cell behaviour patterns that have been observed appear likely to play important roles in both these processes. One can explain most of the morphological events in terms of contractions of microfilaments, expansion of microtubules, differential cell affinities and different cell behaviour. What we cannot yet explain is how the events *start*.

8

Interactions between Tissues in the Early Embryo

Ever since microsurgical and tissue culture techniques were introduced into embryological work, making it possible for pieces of embryos to be transplanted to new sites, or grown in isolation, it has become well known that the course of differentiation in many early tissues and organs is controlled by other tissues with which they are in contact during their development. Generally, a tissue grown in isolation is less capable of developing normally than it is when left in contact with its usual neighbouring tissues. Occasionally, though, it may do better and form more than its normal range of structures. The dependence of neighbouring tissues is often mutual: when a piece has been removed, the remaining areas may not be able to develop normally. So, while it is still right to emphasize, as we have done in Chapter 6 (p. 133), that embryonic cells are remarkably adaptable and are often able to make good any losses of parts and to form a normal embryo with the remainder, this does not always hold true when the tissues and organs have already begun to form. At these stages there are continual interactions between the tissues, which are essential to their normal differentiation. There are limits to how well the embryo can adapt to losses of material, according to whether or not certain 'key' tissues on which many others are dependent for their normal development are left intact. One of these 'key' tissues at the early gastrula stage in amphibians is the dorsal mesoderm which invaginates under the dorsal lip of the blastopore. The famous grafting work by Spemann and his colleagues during the period from about 1916 to 1925 demonstrated that this dorsal lip mesoderm exerted an essential, 'inducing' action on the ectoderm under which it came to lie as it invaginated, and caused the ectoderm to differentiate into neural tissue (see Spemann, 1924, 1938). If the mesoderm was removed, or was prevented from invaginating, no neural differentiation would occur in the dorsal ectoderm. More demonstrative still, if the dorsal lip was grafted onto another gastrula, it would induce the host ectoderm overlying it to form an extra nervous system and other axial tissues with it.

As a result, a complete extra embryo might develop, giving the host the appearance of conjoint twins. These spectacular results gave a tremendous impetus to experimental embryology in the 1930's, and it was soon found that the equivalent mesoderm regions in other vertebrates had similar abilities to induce neural tissue. The primitive streak of a bird was even found to be capable of inducing the ectoderm of a rabbit gastrula to neuralize, and *vice versa*: in other words, neural induction was by no means a species-specific process. It was later shown that very non-specific, mechanical or chemical agents could also in certain circumstances bring about neural differentiation in amphibian gastrula ectoderm. The earlier work in this field has been reviewed in several text books (e.g. Balinsky, 1970; Saxèn & Toivönen, 1962).

There are some special features to be noted about neural induction in *Xenopus*. As we have already stressed, the dorsal lip of *Xenopus*, unlike that of other amphibians so far studied, is lined by endoderm and not mesoderm. So when this dorsal lip is used for transplants, it is likely that much of the anterior mesoderm will not be included, for it will already have passed forwards. This explains why Holtfreter (1955) reported a prevalence of inductions of caudal tissue, rather than brain, by the dorsal lip of this species. It has also to be borne in mind that at the earliest stage recognizable as a gastrula, when the first signs of the dorsal lip appear, events in the *Xenopus* embryo are further on than in other species, since the already-progressing dorsal mesoderm has already had a chance to initiate some neural differentiation. Neural induction in *Xenopus* must therefore begin a little before there are external signs of inward movement of dorsal tissues.

Another point worth stressing is that although neural induction was the first event to catch the interest of experimenters, it may not be the first tissue-interaction in the early *Xenopus* embryo. Sudarawati and Nieuwkoop (1971) have shown by cell-culture experiments that in *Xenopus* the mesoderm originates by an interaction between cells that seem to be ectoderm and endoderm, at the blastula stage. It is not yet known whether this finding is true of other amphibians too.

Neural induction has been for many years the most extensively-studied of the tissue interactions in the gastrula. We shall now consider briefly some of the approaches that have been used recently in efforts to discover the nature of the stimulus that passes from mesoderm to ectoderm in amphibians. *Xenopus* has been used for much of the latest work in this field. During the last two decades, a number of purified extracts of foreign tissues have been tested for their inducing ability, by placing coagulates of them in contact with amphibian gastrula ectoderm. Besides this approach, attempts have been made to detect the passage of substances from mesoderm to ectoderm in the intact embryo, and to study the early biochemical changes in the induced ectoderm. None of these approaches has yet led to a discovery of '*the*' inductor, though they have given indications of what kind of substance may be involved.

1. Inductors Extracted from Foreign Tissues

Biochemical methods for extracting proteins and nucleic acids advanced rapidly in the 1940's. Trials of a wide variety of whole tissues and crude extracts in the 1930's had already indicated that tissues rich in RNA worked best as neural inductors: so the biochemical embryologists who resumed work after World War II used RNA-rich tissues as their first sources of purified extracts. In Japan, Yamada and his co-workers (see Yamada, 1958) used mammalian liver and kidney, while in Finland, Toivönen and his colleagues used guinea-pig bone marrow (see Saxen & Toivönen, 1962). In Germany, the team led by Tiedemann worked with something a little less 'foreign', namely homogenates of whole 9-day chick embryos. This RNA-rich 'embryo extract' is commonly used as a nutrient and growth stimulator for tissue cultures of various kinds.

It is Tiedemann's group who have made most progress recently in isolating a series of ribonucleoprotein fractions which have proved to be effective neural inductors. Among their extracts they have also found some components which induce ectoderm to form axial mesoderm instead of neural tissue. This is not so surprising when it is realized that normally some cells at the caudal end of the neural plate in amphibians undergo a secondary transformation into axial mesoderm when the tail forms (Spofford, 1948). Tiedemann's team have shown by further fractionation work, using chromatography on DEAE cellulose and CM-cellulose, that neural inductors are eluted at pH 5·6, while mesoderm inductors, which are proteins other than ribonucleoproteins, elute at pH 7. However, further analysis of each apparently 'pure' neural or mesodermal inductor has often yielded fractions with various grades of inducing ability: there seems to be an enormous spectrum obtainable. One really needs a more precise test of inducing ability, instead of the classic 'implant' or 'explant' in which either a coagulum is inserted into the blastocoel of a gastrula, or it is placed in between two isolated pieces of gastrula ectoderm, or the ectoderm is grown in a medium to which the inductor has been added in solution. There are many imprecisions in tests of this kind. For instance, Barth and Barth (1962) showed that even minor changes of ion concentration in a culture medium can affect the differentiation of embryonic ectoderm. Others (Jones & Elsdale, 1963; Deuchar 1970a, b) have shown that the ability to differentiate is influenced by the number of cells originally present. Since during any period of culture the cell numbers and the composition of the medium may change owing to cell death and to accumulation of waste products, the conditions are not stable. Cultures of *Xenopus* ectoderm cells need not be kept so long as those of other amphibians, as they differentiate relatively rapidly (cf. Chapter 5), but their metabolic rate is probably correspondingly higher, so accumulation of waste products is just as much of a problem here as with cells of other species.

From the results with foreign inductors reported so far, it seems that ribonucleoprotein is the most likely material to be the normal inductor *in vivo*. But to confirm this, we need evidence that a ribonucleoprotein *present in dorsal mesoderm*, rather than extracted from foreign tissues, will work as a neural inductor. Unfortunately it has not so far been possible to carry out the long fractionation procedures used by Tiedemann's team, on dorsal lip extracts, simply because these provide such small quantities of material. But some progress towards this goal has been made. Kocher-Becker and Tiedemann (1968) obtained RNA extracts from whole embryos of the newt, Triturus, pooling several developmental stages, and found that these gave weak neural inductions. I made some crude extracts of different tissues of *Xenopus* gastrulae in sucrose-saline, and found that the extracts from dorsal lip tissue gave much higher percentages of neural inductions than did extracts from other gastrula regions (Deuchar, 1967). Recently Faulhaber and Geithe (1972) have gone much further than this in analysing extracts of whole *Xenopus* gastrulae. Using hydroxyapatite columns, they have separated protein, RNA, RN-protein and deoxy-RN-protein extracts and have tested these separately on amphibian ectoderm. They find that certain protein and high-molecular-weight RN-protein fractions induce hindbrain and spinal cord, with some mesoderm, while the DN-protein components are only very weak inductors. Faulhaber (1972) has also analysed the effects of subcellular components from homogenates of gastrulae, and has found that microsomal fractions give mainly forebrain inductions, while all other fractions, including yolk platelets, induce hindbrain and spinal cord. So far these workers have not used extracts from the dorsal lip alone, however, so we do not know yet whether there is some additional, perhaps quite different component at work in this normal neural inductor.

Even if purified components are eventually extracted from dorsal mesoderm tissue and found to bring about neural induction in gastrula ectoderm, it will still be necessary to show that materials of this kind normally pass from mesoderm to ectoderm in the intact gastrula, and are the normal inducing agents. We should now consider the observations that have so far been made, many of them on *Xenopus* gastrulae, of the extent to which materials pass from inductor to induced tissue.

2. Passage of Materials from Inductors into Ectoderm

In order to trace a material from one tissue to another, it needs to be 'labelled' in some way. For this purpose radioactive tracers which can be detected by autoradiography or scintillation counting were formerly used.

But as most of the labelled materials available were small molecules—
e.g. amino-acids or nucleotides—their passage into ectoderm which had been
put in contact with mesoderm containing the tracer would not prove that
whole proteins or nucleic acids had passed across intact. So since the time
that it became evident that the neural inductor was probably a macro-
molecule, labelling of whole proteins with fluorescent antibodies (Coons &
Kaplan, 1950) has been the preferred technique. Among the pioneers of such
methods for studies of neural induction, Vainio, Saxèn and Toivönen (1960)
showed that fluorescently-labelled antigenic material, presumed to be whole
proteins, passed from bone marrow into ectoderm which had become
neuralized in contact with it. However, a little labelled material also passed
across in cases where the ectoderm did not neuralize, so there was some
doubt as to whether this material played any essential part in neural induction.

It has been possible to obtain some idea of the *size* of molecule which must
pass from mesoderm to ectoderm if induction is to occur, by work in which
filters of known pore size are inserted between the two tissues in an intact
gastrula. Thus, Brahma (1958) showed that in *Xenopus* Gradocol membranes
of pore sizes of 4 mμ or smaller prevented neural differentiation. One could
interpret this either as meaning that a large molecule, or that a very rapid
passage of small molecules is necessary for induction to occur. Since Nyholm
et al. (1962) showed that filters as thick as 20–25 microns do not prevent
induction in *Xenopus*, it seems most likely that small, rapidly moving
molecules are concerned. Incidentally, examination of filters with the
electron microscope has shown that there is no penetration of cytoplasm
across them during induction, thus confirming that there is no need for
contact between the mesoderm and the ectoderm.

Since the advent of electron microscopy, it has been possible to study in
detail the structural elements in the interzone between mesoderm and
ectoderm. Both Kelley (1969) and Tarin (1971b, 1973) aroused interest by
reporting the presence of granules in this interzone, many of which disappeared on treatment of the tissues with RN-ase. However, similar granules
were also seen between ectoderm and endoderm (Tarin, *loc. cit.*), so it has
become doubtful whether they represent any specific interaction between the
tissues. It has so far not been possible to show where the granules originate
from, or what is their final fate. Kelley (*loc. cit.*) observed lobules projecting
from the surfaces of the mesoderm cells in the regions where induction
should be taking place, and projections similar to these have also been seen
to be rich in granules of the same kind as those that lie in the interspace
between mesoderm and ectoderm. But this is no proof, without further
evidence, that the granules originate from the mesoderm. Yet another
enigmatic feature of the granules has come to light: many of them respond to
PAS (Periodic Acid–Schiff) stain and are dissolved by amylase, indicating
that they consist mainly of glycogen rather than RNA. This makes it less

plausible that they have anything in common with other tissue extracts that are successful neural inductors. As we saw in Chapter 6, many cells at earlier stages have been seen to shed glycogen into the blastocoel, so the dorsal mesoderm is continuing a process that has been going on since cleavage stages. The only role we know of for glycogen in animal cells is as an energy source. It is certainly likely that extra energy is required for the process of neurulation, but glycogen is not the kind of molecule that one would expect to initiate the differentiation of these cells. It is not known to have any effects on gene activity, whereas RNA could be directly implicated in gene action. It is interesting to recall that some time ago Brachet claimed to have seen RNA-rich material passing from mesoderm to neural plate in the amphibian gastrula. His observations were based on light microscopy.

3. Observations on the Changes that take place in Induced Ectoderm

Electron microscopy has been applied also to studies of the immediate structural changes that occur in ectoderm after it has been subjected to a neural inductor. Eakin and Lehmann (1957) pioneered in such work, and despite imperfect fixation methods at that time, managed to show that the neural ectoderm acquired denser endoplasmic reticulum (an assembly of ribosomes on membranes, characteristic of the onset of protein synthesis). They also observed that the mitochondria of the ectoderm became elongated, with more cristae than in other gastrula tissues. It is surprising that these studies have not been repeated by other workers now that fixation methods have improved. But recently investigators seem to have ignored the ultra-structure of the induced ectoderm while concentrating on the ectoderm/mesoderm interface before and during the induction period.

A number of biochemical changes have been observed in neural ectoderm immediately after induction. Interest has so far been focussed mainly on proteins and RNA. Clayton (1953) established that the neural plate ectoderm of the early neurula in Triturus has antigens which differ from those in other tissues, including epidermal ectoderm. More recently, Stanisstreet and Deuchar (1972) have been able to show by methods that combined immuno-logical and electrophoretic separation of the antigens, that three antigenic components become prominent in the neural ectoderm by the end of gastrulation, immediately after it has received the induction stimulus and before it has begun to differentiate visibly into neural plate (Figure 8.1).

Preceding protein changes, one would expect to see changes in RNA: particularly, the appearance of new DNA-like RNA and also of polysomes to which nascent protein is attached. It has been very difficult to make precise studies of such changes, because they represent such small proportions of the total RNA. In order to get enough material to analyse, large numbers of pieces of ectoderm have to be dissected from embryos in the short time

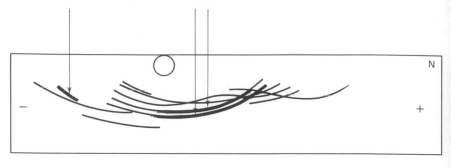

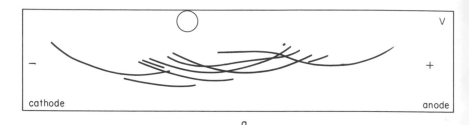

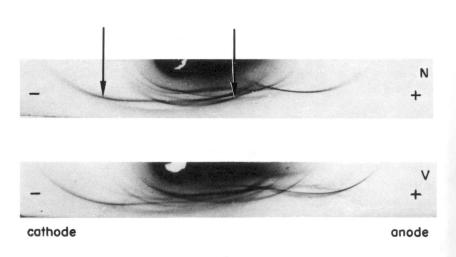

Figure 8.1. Antigens in late gastrula ectoderm of *Xenopus*, separated by immuno-electrophoresis. N = neural ectoderm; V = ventral ectoderm. (a) Diagram of the antigenic bands seen:—a composite of all the clearest runs. (b) Photograph of one result. Arrows indicate antigenic material present in higher concentration in the neural ectoderm. (From Stanisstreet & Deuchar, 1972)

between the end of gastrulation and the beginning of neurulation. This is an even shorter period in *Xenopus* than in other more slowly-developing species (cf. Chapter 5). However, Waddington and Perkowska (1965) used embryonic tissues of *Triturus* and obtained profiles of RNA extracted from these, separated into various molecular sizes by centrifugation on a sucrose density gradient. They found some slight differences between ectoderm and other tissues, at the neurula stage (Figure 8.2 (a)). Later, Evans (1969) managed to obtain samples of polysomes extracted from dorsal and ventral ectoderm of late gastrulae of *Xenopus*. He found two classes of polysome in each type of ectoderm that did not appear in the other type, nor in early gastrula ectoderm. Thomas and Deuchar (1971) analysed extracts of radioactively labelled (i.e. newly synthesized) RNA from dorsal and ventral late gastrula ectoderm of *Xenopus*, using acrylamide gel electrophoresis to separate the different molecular sizes. We found that heterogeneous, high-molecular-weight material, possibly d-RNA, was synthesized in slightly higher concentrations in the dorsal, induced ectoderm than in the ventral, uninduced ectoderm (Figure 8.2 (b)).

4. Blockage of Neural Induction by External Agents

Despite the variety of findings just described, it is still not possible to define what exact biochemical process is responsible for neural induction. But according to current ideas, any process of differentiation must at some stage involve gene action and the liberation of d-RNA into the cytoplasm, where it becomes attached to ribosomes which control the sequences of amino-acids in newly synthesized proteins. So to be effective in starting a process of differentiation, any 'inductor', neural or otherwise, must be able to initiate one or more of these steps in gene activity. We have just seen in the last section that there is plenty of evidence of increased gene activity, with increased d-RNA production and the appearance of new proteins, in the ectoderm of amphibians immediately after exposure to inducing mesoderm. To find out whether this activity is essential to neural differentiation, however, it would be necessary to block it experimentally and then to see if neuralization was also blocked. A suitable agent for blocking gene transcription is Actinomycin D, which combines with guanine in the DNA molecule and thus blocks its transcription into RNA. Denis (1964) used this agent on explants of dorsal lips combined with gastrula ectoderm. Combining either dorsal lip which had been treated with Actinomycin, with normal ectoderm, or normal dorsal lips with Actinomycin-treated ectoderm, prevented neural differentiation. So evidently *both* the inducing powers of the lip, *and* the ability of the ectoderm to respond, were blocked if gene transcription was prevented. If the inducing agent is some kind of protein, it is of course to be expected that gene transcription would be necessary for its

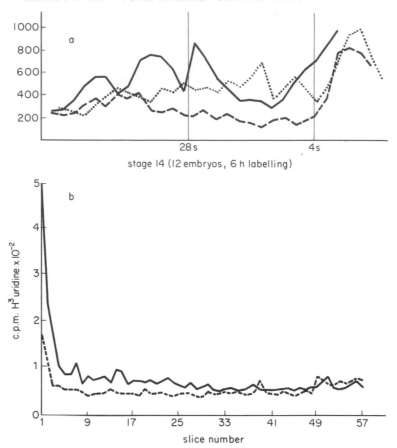

Figure 8.2. Differences in the RNA extracted from neural and from ventral ectoderm in amphibian embryos. (a) above: RNA from axial tissue (_____), ectoderm (. . . . . .) and endoderm (----------) of neurulae of *Triturus*, from Waddington & Perkowska, 1965. (b) below: RNA extracted from neural ectoderm (_____) and ventral ectoderm (------) of late gastrulae of *Xenopus laevis* (Thomas & Deuchar, 1971). Note that in (b) there is a higher rate of RNA synthesis, indicated by higher radioactivity (c.p.m.) in neural ectoderm

production. This work therefore supports the indications from the tests with foreign tissue extracts, that proteins, especially ribonucleoproteins, are the most active inductors.

There are other factors which govern the success of neural induction in the normal, intact gastrula. One is the extent to which the mesoderm invaginates, and the other is the extent to which adequate numbers of ectoderm cells are able to interact after being induced. Experimental procedures which limit either of these processes will result in poor or absent neural differentiation.

Among several treatments which have been found to block invagination, lithium chloride has been used most effectively. Backström (1954) found that low doses of this agent produced damage to the notochord and stunting of the neural axis in *Xenopus* embryos. Excess magnesium ions had similar, but not so marked effects (Backström 1962). The stunting was accompanied by reduction in size of the brain and spinal cord, and sometimes by absence of parts of the forebrain. Another class of compounds which inhibit invagination and therefore also reduce the efficacy of neural induction are the thiols. β-mercaptoethanol has been used on *Xenopus* embryos by Tuft (1961), and Brachet (1961) used lipoic acid and dithioglycol. Both of these agents caused nervous system defects, including microcephaly. All these results underline the need for an adequate area of ectoderm to be underlain by mesoderm, if the induction process is to be normal—even though, as we have seen earlier, induction can occur without direct contact between the two layers.

Experimental procedures in which very small numbers of ectoderm cells have been tested for their powers of neural differentiation (Jones & Elsdale, 1963; Deuchar, 1970a, b) have shown that a full range of neural tissue will not form unless there are adequate initial numbers of ectoderm cells in the presence of the inductor. Fewer than 40 cells will not differentiate at all, and unless a minimum of 100 cells are present initially, the differentiation is unstable and tends to regress after a few days (Deuchar, 1970a). It is, however, possible to obtain neural tissue from very small numbers of induced ectoderm cells if these are combined with a larger piece of uninduced ectoderm. This shows that the cells are able to pass on the induction stimulus in some way to the larger piece. Using tritiated thymidine to label the induced cells. it has been shown that they co-opt uninduced cells from the larger piece, to participate in forming the neural tissue (Deuchar, 1971; see Figure 8.3). These findings suggest that neural induction is a two-stage process: first, transmission of a stimulus from mesoderm to ectoderm, then, transmission of this (or perhaps another) stimulus between the ectoderm cells, so that they co-operate to form a morphologically normal nervous system. These two steps may be equivalent to the 'evocation' and 'individuation' described by earlier workers (see Waddington, 1952).

5. Interaction during Further Differentiation of the Embryonic Axis

Experiments on birds as well as amphibians have shown that each region of the central nervous system and other axial tissues is influenced in its development by the regions adjacent to it cranially and caudally. This is most easily demonstrated by experiments in which segments of the axis are either removed, or their cranio-caudal level altered. Removal of small pieces of the neural plate, for instance, results in no abnormality of the central nervous system because the defect is made good by extra growth of the regions

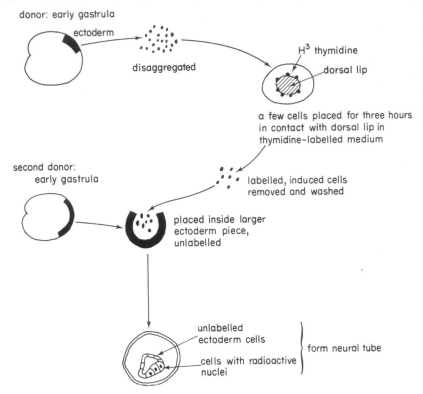

Figure 8.3. Diagram of labelling experiments to show that an induction stimulus can be transferred from small numbers of ectoderm cells to a larger ectoderm piece. (Based on Deuchar, 1971)

cranial and caudal to the wound (Corner, 1963, 1964). These regions have evidently responded to a change in their immediate neighbourhood, and it is to be assumed that normally their growth keeps pace with that of adjacent regions, in the unoperated embryo. This was shown to be the case, by experiments on *Xenopus* in which lengths of the neural plate or neural tube were rotated 180° (Hunt, 1969). This operation brings the narrower, caudal edge of the piece into contact with the broad, cranial end of the neural plate remaining, while the broader end of the piece comes in contact with the narrower, caudal region of the plate. It is found, however, that if the operation is done on the early neurula, growth adjusts at these borders, to make a smooth join and in the end a symmetrical spinal cord of normal size. Operations at later stages result in less complete adaptation of size, with a tendency for cranial regions of either graft or host to adapt less readily than caudal regions. Thus it looks as if cranial regions dominate in the size adjustments that must occur in normal differentiation of the embryonic axis.

The somites provide an example of even more precise control of differentiation along the cranio-caudal axis. The mechanism by which a constant number of somite pairs is formed in each animal species has always intrigued and baffled investigators. Adding or subtracting tissue from the somite mesoderm of a gastrula makes no difference to the eventual number of somites, but will either increase or decrease their individual sizes (Waddington & Deuchar, 1953). Evidently there is some 'counting' mechanism in the mesoderm which causes it to continue segmenting until the correct number of somites is formed, then 'tells' it when to stop. Mathematicians and geneticists are still grappling with possible models for the control of the development of serially-repeated structures, but no model has yet been devised which can be applied satisfactorily to all the events of somite segmentation. For readers interested in this topic, discussions will be found in Turing (1952), Maynard-Smith (1960) and Goodwin (1971).

There is controversial evidence from work on chick embryos that some 'signal' passes cranio-caudally through the somite mesoderm, particularly from the node region. Posterior parts isolated from the node by transverse cuts are unable to segment. This is thought by some workers (Spratt, 1957; Nicolet, 1970) to be because the posterior regions lack some stimulus from the node region, especially since when a node is transplanted into these pieces they are then able to segment. However, Bellairs (1963) disagrees with this view and regards the regression process undergone by the node after gastrulation as the important morphogenetic influence in these explants. In *Xenopus* and axolotl embryos, Burgess and I (Deuchar & Burgess, 1967) were unable to find any evidence of an influence passing cranio-caudally in the somite mesoderm. We did various operations in which this mesoderm was interrupted, either by deleting or inverting short lengths of it, or by bisecting the whole embryo (Figure 8.4). None of these operations prevented the segmentation of somites in posterior parts. So we concluded that either the signal must have passed along at a stage prior to the operations, or the individual regions of the somite mesoderm must segment autonomously.

A special word needs to be said about the mode of formation of segmental muscles (myotomes) in *Xenopus*, since these show peculiarities unlike other amphibian species studied. Hamilton (1969) observed that the myocoel is obliterated before segmentation starts, which is an unusual feature, and also that the final longitudinal orientation of myoblasts is achieved by a *rotation* of the myotome cells through 90°. These cells were originally orientated with their long axes vertical, and the 90° rotation brings them into their final, cranio-caudal orientation. The gradual rotation can be seen in sequence along the axis, since it starts in the most cranial somites and follows on in cranio-caudal sequence through the series.

In the formation of the embryonic axis there are of course many other tissue interactions which there is not space to mention here. These include

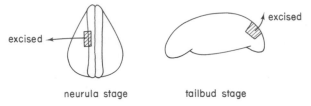

a. deletions of pieces of unsegmented somite mesoderm and neural tissue

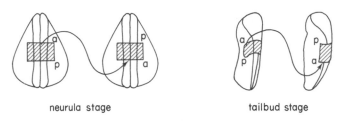

b. antero-posterior inversions

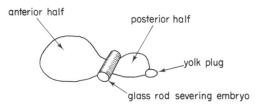

c. transverse bisection; late gastrula stage

Figure 8.4. Experiments to test the possibility that somite segmentation is controlled by a stimulus passing from anterior to posterior in the somite mesoderm. Redrawn from Deuchar and Burgess, 1967. (a) unilateral deletions of somite mesoderm, at neurula or tailbud stages; (b) antero-posterior inversions of neural and underlying somite tissue at neurula and tailbud stages; (c) transverse bisection at late gastrula stage

medio-lateral as well as cranio-caudal interactions. Just one example is that the notochord is essential to the normal development of somites, and the somites in turn influence the development of lateral mesoderm. Ranzi and Gavarosi (1959) found that treatment of *Xenopus* embryos with thiocyanate tended to enlarge the notochord, and that it contained more cells than normal. Evidently in compensation for this, there were fewer cells in the somites, showing that they had responded to the alteration in the tissue medial to them, just as parts of the neural plate were shown in Corner's and Hunt's experiments to adapt to changes of tissues cranially and caudally to them.

Reviewing the topics covered in this chapter, it can be seen that certain features of tissue interaction in early embryo have been elucidated by work particularly on *Xenopus*. The inducing properties of extracts from the dorsal lip have been tested, in *Xenopus* material; the fine structural components lying between mesoderm and ectoderm have been studied particularly in *Xenopus*; and the biochemical changes in the ectoderm immediately after induction have been shown up in this species. *Xenopus* has also been used for experiments in which the interactions within ectoderm, and within somitic mesoderm during segmentation, have been studied in more detail. *Xenopus* has been found to differ from other amphibians in the mode of formation of the myotomes, and in the limited inducing power of its dorsal lip, which is explicable by the absence from it of anterior mesoderm.

9

Organogenesis in *Xenopus*

It is not intended to give here a detailed description of how all the organs are formed in *Xenopus*. For those who need such information, the best source of references is the handbook by Nieuwkoop and Faber (1956) which has already been mentioned. The present chapter will deal with just a few recent descriptive and experimental discoveries about organ and tissue development in *Xenopus* which are of interest when compared or contrasted with events in other vertebrate embryos.

We must first consider some of the events that occur at the biochemical level during early differentiation in the tissue primordia. Like all other steps in embryonic differentiation, this involves the onset of new gene activities, which differ in different groups of cells. We saw in the last chapter that some evidence of these different gene activities begins to show already at gastrula and neurula stages. At subsequent stages and at the onset of organogenesis too, one finds evidence of changing gene activities, showing up as alterations in the newly-synthesized RNA and proteins in the cells.

Earlier work on the distribution of new types of RNA and protein in tissue primordia was carried out on the newt, *Triturus* and the frog, *Rana*, but in the last decade *Xenopus* has become preferred material, because it can be induced to ovulate at any time of year and produces large numbers of eggs at each spawning, so it is particularly useful for biochemical work where regular and abundant supplies of material are essential. The RNA and DNA of *Xenopus* embryos has now become better characterized than that of any other amphibian species: hence the tendency is for people to continue using this well-known material. Also the discovery of the anucleolate mutant in *Xenopus*, which has little or no ribosomal DNA or RNA, has made it easy to recognize r-DNA and r-RNA in biochemical preparations, by comparisons between anucleolate and normal embryos.

In the following account we shall consider first RNA synthesis and then protein synthesis during the differentiation of tissue primordia.

1. New RNA in Tissue Primordia

As was already stressed in Chapter 4, there is no major RNA synthesis in *Xenopus* embryos until the onset of gastrulation. Maternal gene products evidently suffice for all RNA-controlled events occurring before this stage. The same is probably true in other amphibians, since Johnson (1969) and Moore (1962) found that inter-species hybrids of *Rana* develop normally up to the gastrula stage, with cleavage rates typical of the maternal species. Most of them then fail to gastrulate (cf. Chapter 7), so it seems likely that this is the stage when the paternal genes come into action. In fact it is possible that none of the nuclei derived from the zygote is directly concerned in controlling development prior to gastrulation, and that the cytoplasm already contains all the products necessary to control cleavage processes. It was shown some time ago by Briggs, Green and King (1951) that even enucleated eggs of *Rana* will undergo a cytoplasmic cleavage and form something resembling a blastula, with smaller cells at the animal pole and larger, more yolky cells in the vegetal region. Differences in the distribution of cytoplasmic particles in the cleavage cells, whether or not these are nucleated, result simply from the unequal distribution in the original egg cytoplasm: they need not be attributed to differential gene activity.

There is, however, evidence that some gene activity begins, on a small scale, at pre-gastrulation stages in *Xenopus*. Brown and Littna (1964) showed that a little d-RNA is synthesized during late cleavage, and Crippa and Gross (1969) have shown that some of the d-RNA which has been present since oogenesis begins to disappear at the mid-blastula stage and is slowly replaced by new d-RNA. These authors used molecular hybridization methods to show that by the time gastrulation starts in *Xenopus*, half of the maternal d-RNA base sequences have been lost. Other workers (e.g. Brown & Littna, 1964; Landesman & Gross, 1969) have shown also that precursors of ribosomal RNA begin to appear a little before gastrulation, at the same time as the nucleoli reappear (cf. Chapter 4).

The distribution of newly-synthesized RNA in different tissue primordia of early embryos is most conveniently studied in histological sections. If radioactive precursors of RNA (e.g. tritium-labelled uridine) are either injected into the embryos or introduced into the medium round them, the precursors will be taken up most rapidly into those regions that are synthesizing RNA most actively. Using autoradiography, the distribution of the tracer, and hence the relative rates of RNA synthesis in different regions, can be seen. By this method Bachvarova, Davidson, Allfrey and Mirsky (1966) were able to show that the dorsal lip mesoderm and prechordal plate of *Xenopus* gastrulae had higher rates of RNA synthesis than other regions. Woodland and Gurdon (1968a) went further and showed by biochemical

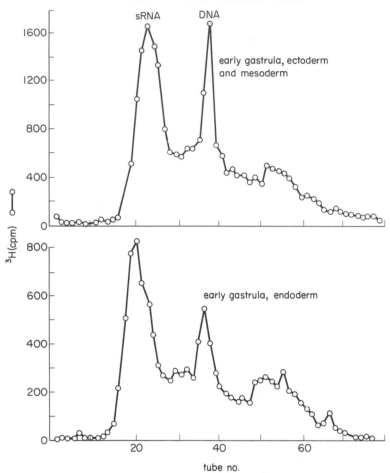

Figure 9.1. Profiles of nucleic acid components labelled by tritiated uridine, in parts of early gastrulae of *Xenopus*. Separation on MAK columns. After Woodland and Gurdon, 1968. It will be seen that the endoderm has a higher ratio of labelled RNA to labelled DNA, than other regions, indicating that it synthesizes more RNA per cell. (At neurula and later stages, the endoderm also synthesizes greater quantities per cell of rRNA than do other regions)

extraction methods that more RNA per cell was being synthesized in ventral endoderm than in other parts of the embryo, at stages from gastrula to hatching (Figure 9.1). These two sets of findings indicate that gene activity is highest both in those cells that invaginate earliest at gastrulation and in cells with large cytoplasmic reserves. These reserves include precursors of DNA, RNA and protein, all of which have been found both in yolk platelets and in the ambient cytoplasm of amphibian embryonic cells.

There is little data available on any differences in the base sequences of RNA synthesized in different tissues or organs of *Xenopus*. It is difficult to get enough material from early embryos for base sequence analyses. At later stages, though, Ford and Southern (1973) have shown that the ovary produces three base sequences which are not present in kidney tissue. Further evidence of this kind would be of great interest, since it indicates that different sequences of DNA bases are being transcribed in the different organs: i.e. that different genes are active in them.

2. Protein Metabolism in Tissue Primordia

(a) The Sources of Raw Materials

Rates of synthesis of specific proteins in the early tissues of yolky embryos depend partly upon the prior mobilization of reserve stores from the yolk. As a result of the different rates at which yolk is broken down in different regions (Selman & Pawsey, 1965) the concentrations and types of free amino acids available for protein synthesis in these regions differ. When different regions and tissue primordia of *Xenopus* embryos were compared, at stages ranging from the blastula to the month-old larva (Deuchar, 1956) some striking differences in their free amino-acid contents were seen (Figure 9.2). It was specially interesting that, from a very early stage, the central nervous tissue showed high concentrations of free glutamic acid and glutamine. This is also a feature of adult brain tissue (Ansell & Richter, 1954). In other dorsal tissues, a high concentration of free leucine was observed, at a stage just before that when leucine is taken up into somite cells, where it will constitute about 10 % of the myosin molecule when this is formed. Another general point of interest is that the dorsal regions, where yolk breakdown occurs earliest, showed the highest total free amino acid. A study of protease (cathepsin) activity in early tissue primordia at these same stages (Deuchar, 1958) confirmed that proteolysis was more active in the dorsal regions. Furthermore, a considerable proportion of the catheptic activity was found to be bound to the yolk platelets, when these were separated from other cell components by low-speed centrifugation. So the yolk platelets may be to some extent self-autolysing. Even so, they have different rates of breakdown in different regions, so breakdown seems to be controlled by differential gene activities. Certainly when only half the genes are present, as in haploids, yolk-breakdown is much slower (Fox & Hamilton, 1971).

(b) The Synthetic Steps

The first step towards assembly of amino-acids onto the ribosomes to form polypeptides is the so-called 'activation' of the amino-acids by

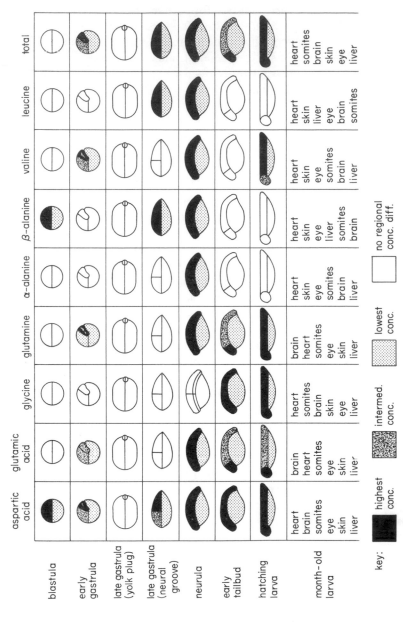

Figure 9.2. Diagrams comparing the concentrations of individual free amino acids in different regions of *Xenopus* embryos and larvae. From Deuchar, 1956

incorporation of adenosine monophosphate onto one of their end-groups. They are then conveyed to the ribosomes by low-molecular-weight RNA called transfer RNA (t-RNA), to which they are attached with the help of enzymes called amino-acyl transferases. Attempts have been made to investigate some of these steps, in embryonic tissues of *Xenopus*. Studies of amino-acid activation (Deuchar, 1961, 1962) showed some quantitative differences between different regions of the embryo: for instance, in somite tissue there was a higher rate of leucine-activation than in other tissues of the neurula. The same was found to be true in chick embryos of an equivalent stage. The types of t-RNA that can be extracted from different embryonic and adult tissues of *Xenopus* have been analysed using RNA-DNA hybridization methods, by Marshall and Nirenberg (1969). They were also able to prepare t-RNA and amino-acyl transferases from the different tissues, then to attach a different radioactive amino acid to each t-RNA. The uptake of radioactivity from these samples into ribosomes was then compared in the presence of several different nucleotides. It was thus possible to show that t-RNA from neurulae resembled t-RNA from adult liver in its ability to recognize certain triplets of bases in the nucleotides. Further work along these lines is needed, before it can be said whether or not the early tissue primordia of the embryo have differences in their t-RNA that can be correlated with their synthesis of different, tissue-specific proteins.

The rates of incorporation of amino-acids into protein in the different regions of embryos have been studied partly by autoradiographic and partly by biochemical methods. For instance, Waddington and Sirlin (1964) showed that 14-C-glycine was incorporated most rapidly into epidermis, then at decreasing rates into somite mesoderm, neural tube, notochord and endoderm respectively, in *Xenopus* embryos at the tailbud stage. Since glycine is not only a very widely-distributed amino-acid but also undergoes many exchange reactions in both amino-acid and carbohydrate metabolism, these findings give only an approximate indication of rates of protein synthesis in the tailbud tissues. A less ubiquitous amino-acid like leucine may give slightly more information: for instance one can show that it is taken up mostly into the proteins of somite mesoderm, at stages when leucine-activation is most marked in this tissue (Deuchar, 1960). Proline, on the other hand, is taken up most rapidly into the notochord and the dermis (Deuchar, 1963), which are two regions where collagen, containing high proportions of proline in hydroxylated form, is beginning to be synthesized by late neurula stages (Green *et al.*, 1968). The proline uptake suggests that some time before collagen synthesis begins, these particular tissues are accumulating precursors of it. I was able to confirm by extracting collagen from embryos at later stages, that the radioactivity from the proline had passed into the hydroxyproline of collagen, which was identifiable by paper chromatography.

(c) Tissue-specific Proteins

There are relatively few tissues which have a monopoly of one particular protein or polypeptide, enabling the first steps in the tissue's differentiation to be detected by the appearance of precursors to this protein. Collagen is specific to connective tissue, but this is very widespread in the body. Other examples are actomyosin in muscle, haemoglobin in red blood cells, and chondroitin sulphate (a polypeptide-linked polysaccharide) in cartilage. Radioactive labelling methods have been used in attempts to trace precursors of the latter two substances. For instance, Pantelouris, Knox and Wallace (1963) followed the localization of radioactive iron (^{59}Fe) in early *Xenopus* embryos, finding it mostly in cells of lateral and ventral regions in late gastrulae and neurulae. These are the regions from which blood stem cells arise, broadly speaking, but unfortunately it was not possible to extend the observations to trace stages of erythropoiesis. Okada and Sirlin (1960) were more successful with radioactive sulphur (35-S) as an indicator of chondroitin sulphate synthesis. They found that it was taken up preferentially into cartilage-forming cells in *Xenopus* embryos, and it could be traced into the matrix of the cartilage eventually.

Some whole proteins are readily identifiable by their enzyme activity: but it is rare for an enzyme to be confined to one tissue, so that most early tissue primordia in the embryo can be expected to show only quantitative and not qualitative differences in enzyme activities, as evidence of their early differentiation. One can show, for instance, that those regions of the gastrula which are invaginating most actively, like the dorsal and ventral lips of the blastopore, have a high glycolysis rate. This was recently demonstrated in *Xenopus* by Waldvogel (1965), having been shown long ago in Triturus by Woerdeman (1933) and in Ambystoma by Jaeger (1945). It seems most likely that the glycolysis provides extra energy for invagination movements. Other energy-producing reactions may also be involved: for instance Weber and Boell (1955, 1962) have demonstrated mitochondrial ATP-ase and cytochrome oxidase activities in *Xenopus* embryos.

One enzyme which has aroused interest recently, since it has a restricted distribution, is monoamine oxidase. It is found chiefly in nervous tissue, where it breaks down 5-hydroxytryptamine (5-HT), an indole compound which acts as a transmitter in the nervous system. Baker (1966) found a steep increase in the activity of this enzyme in post-hatching stages in *Xenopus*, and noted that most of the activity was in brain and eye tissue. Further work may establish how early in pre-hatching stages this enzyme is detectable in the central nervous system.

Although the majority of easily-detectable enzymes are ones which take part in essential metabolic processes and are therefore ubiquitous in all tissues, many of them are now known to have several different molecular forms, differing in their polypeptide subunits. Each tissue has different, characteristic forms of the enzymes. The various forms are called 'iso-enzymes'. The best-known example is lactate dehydrogenase, which has four main subunits. The subunits can be identified by electrophoretic methods.

In a number of recent studies on isoenzymes in developing tissues of vertebrates, it has been shown that each tissue during its embryonic develop-ment exhibits a steady trend towards the isoenzyme characteristics of the adult tissue (for reviews, see Markert & Whitt, 1968; Masters & Holmes, 1972). In developing tissues of *Xenopus*, Kunz and Hearn (1967) demon-strated several variants of lactate dehydrogenase. They were able to identify at least nine subunits in the molecule and to show that these changed in their proportions during the development of each tissue (Figure 9.3).

The work on isoenzymes in other animals has confirmed what we now regard as true of gene activity: namely that each genetic unit (or 'cistron', to use the recent term) governs the production of one polypeptide. Many of the gene loci which control the synthesis of subunits of lactate dehydrogenase in the mouse have now been characterized (Ruddle & Harrington, 1967). So it may be that our best hope of detecting differences in gene activity in early tissue primordia of *Xenopus* and other vertebrates is in future to look for changes in readily-identifiable polypeptides such as the subunits of isoenzymes.

3. Recent Work on Individual Organs of *Xenopus*

This will have to be a very selective account and will, I am afraid, omit much good and promising work on organ cultures from *Xenopus*. This has been reviewed fully by Monnickendam and Balls (1973), however. We shall deal here with only a few points of general biological interest that arise from experimental and descriptive work on the development of some of the organs in *Xenopus*.

(a) The Skin and its Components

The skin is composed of several tissue layers, each containing a number of important organs and organelles. We shall say a very few words about recent findings from studies of the skin glands, the lateral line organs and the chromatophores.

Skin Glands. The granular glands of the skin of *Xenopus* were described by Vanable (1964). They are of special interest to neuroendocrinologists because they secrete 5-HT, the transmitter mentioned in section 2. There are also mucous glands, whose development has been followed by Vanable and Mortensen (1966). Each of these glands arises from a *clone* of cells, derived from one single stem cell, and each gland is capable of completing its development *in vitro* if it is isolated at an early stage of its differentiation.

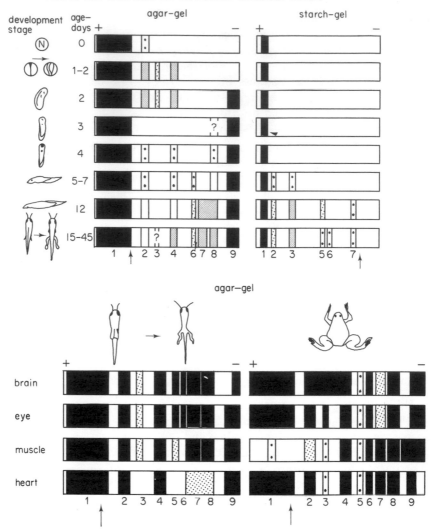

Figure 9.3. Diagrams illustrating the different isozyme subunits of lactate dehydrogenase identified at different stages and in different tissues of *Xenopus laevis*, by electrophoresis on agar-gel and starch-gel. After Kunz and Hearn, 1967

Another interesting feature of these glands is that they respond *in vitro* to thyroxine, which causes them to regress, just as they would do in the intact animal at metamorphosis (McGarry & Vanable, 1969a, b).

Lateral Line Organs. The special interest of these organs in *Xenopus* is that they persist in the adult, instead of being lost at metamorphosis as in other anurans which change from an aquatic to a terrestrial life. Their structure has been studied in some detail in the adult by Dijkgraaf (1962) who described the remarkable strap-shaped cupula which overhangs each organ (Figure 9.4). Because of its greater resistance to deformation by vibrations at right-angles to its surface than to those in the same plane as its surface, the cupula provides a mechanism for distinguishing between vibrations received in the two planes. Shelton (1970) has described the changes which the lateral line organs undergo in *Xenopus* at metamorphosis. The larval organs consist

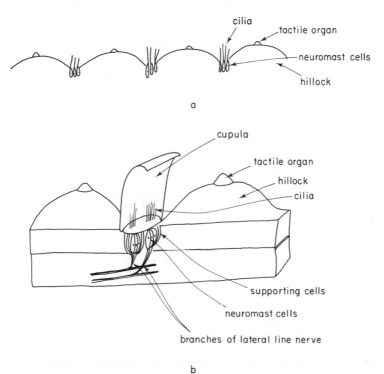

Figure 9.4. The arrangement and structure of the lateral-line organs in *Xenopus*. (a) Sectional view diagram of a row of organs in larva and adult, redrawn after Shelton, 1970. (b) Detail of one organ, redrawn after Dijkgraaf, 1962

of groups of ciliated cells on hillocks on the surface of the skin, but these sink below the surface at metamorphosis leaving the specialized cupula exposed (Figure 9.4 (a)). It has been shown by Murray (1956) that the lateral line organs emit a continuous spontaneous discharge along their sensory nerves. The rate of this discharge is inversely proportional to the environmental temperature and responds to changes of as little as 1–2 °C. Afferent impulses from the lateral line organs can cause central reactions which set off muscle twitching and hence locomotion (Russell, 1971a, b). The main transmitter substance in this system appears to be acetylcholine.

Epidermis.   Other epidermal cells besides those of the lateral line organs are capable of emitting electrical discharges. These discharges have been measured in early larval epidermis of *Xenopus* by Roberts and Stirling (1971). They found that all-or-nothing impulses are passed from cell to cell, and they suggest from electron microscope work that the route of passage is via the tight junctions or 'zonulae occludentes' that are found between these cells (Farquhar & Palade, 1964, 1965). Another kind of epidermal activity is that of the cilia. In explanted pieces of embryonic epidermis, these cilia begin to beat at the equivalent of the neurula stage, causing the explants to gyrate round the culture dish (Deuchar, 1970b). Steinman (1968) has described the development of the cilia from precentriole precursor bodies, visible in electron micrographs. The central axonemal thread of each cilium develops from another body, the 'axonemal precursor body'. We do not know what chain of cause and effect underlies the development of the cilia, since no experimental work has yet been done on this. Steinman postulated that the precentrioles are caused to form by a dense, opaque body which lies at the centre of each group: he called this the 'precentriole organizer body'.

Chromatophores.   These have been studied far more than any other kind of dermal cell, mainly because as soon as they start melanin synthesis they are easily identifiable and can be seen macroscopically as well as microscopically. It was established first in amphibians, that the melanophores are derived from the neural crest (Hörstadius, 1950): this is now known to apply to all other vertebrates too. The exact origin of both melanophores and xanthophores (producing yellow pigment) in each region of the body in *Xenopus* was worked out by Stevens (1954). He isolated segments of neural crest, either *in vitro* or attached to ventral halves of neurulae, and followed their subsequent development. He showed that all chromatophores were derived from the crest in the longitudinal lateral folds of the neural plate, thus refuting the previous ideas that some of the head chromatophores came from the anterior transverse fold region. Stevens also showed that xanthophores would develop only if they were placed either in the peritoneum or in the eye region: two places where they are normally found in *Xenopus*. Grafting the xanthophores elsewhere prevented them from differentiating.

Andrew and Gabie (1969a, b) have described stages in the differentiation of melanoblasts of *Xenopus*, using a hanging-drop method to culture these cells. They noted that as pigment is synthesized, the yolk in the cells is rapidly absorbed and the mitochondria become more prominent. Külemann (1960) suggested that mitochondria provide energy for the movements of pigment within the cells. A peculiarity of the melanophores in *Xenopus* is that the pigment granules in them *disperse*, instead of aggregating, in response to the hormones ACTH and adrenalin (cf. Chapter 3). The pigment response in melanophores of the tail of *Xenopus* larvae is unusual too (Bagnara 1960), for it disperses in bright light and concentrates in darker surroundings. Melanocytes of the epidermis also contribute to colour changes in the intact animal (Hadley & Quevedo, 1967).

There are other puzzling features about the melanophores of *Xenopus*. For instance, MacMillan (1972) has found that their rate of differentiation is more rapid, the denser the clusters they form in culture. Another unexplained phenomenon is that in the normal embryo, clusters of melanophores are often seen in the cavity of the developing brain, where their function, if any, is quite unknown (Adam, 1954). According to Kominck (1961) they migrate to this site from the neural crest, actively penetrating the neuroepithelium of the hindbrain and through other parenchymatous brain tissue to reach its internal cavity. After circulating for some time in the cerebrospinal fluid, they degenerate and are engulfed by phagocytes. It has been reported that melanoblasts themselves have powers of phagocytosis and will engulf their dead fellows in a tissue culture (Niu & Twitty, 1953). However, in Lehman's (1953) observations on tail melanophores in intact *Xenopus* larvae, the phagocytic cells never synthesized melanin—so it is problematical whether they should strictly speaking be regarded as melanoblasts. They also failed to respond to the hormones which cause melanophores to expand.

(b) Special Features of the Gut and Urinogenital System

Because *Xenopus* larvae are filter feeders, the larval gut has some specializations that are not found in other amphibians. We saw in Chapter 5 that a mucous thread is used at first to transport microparticles of food down to the stomach. Dodd (1950) traced this movement, by feeding larvae with colloidal graphite. He found that the particles were packed gradually into a spiral as the mucous thread which entrapped them was drawn into the stomach, by a continuous gulping action. The ciliated tract in the pharynx, described by Weisz (1945) (cf. Chapter 2) entraps particles as small as 0.126 μm. in diameter, according to Wassersug (1972). This means that it is as efficient as the finest man-made mechanical sieves. Wassersug has also shown that the rate of clearance of fluid through the larval pharynx is modified according to the concentration of food in the medium.

In the midgut and hindgut, the further progress of food particles is evidently aided by cilia. Fox, Mahoney and Bailey (1970) found in a series of ultrastructural studies that the gut is ciliated throughout its length. They also emphasized that the cilia in the pharynx are orientated in grooves, which assist the rapid downward passage of the food particles (Fox, Bailey & Mahoney, 1972). There are also cilia at the extreme posterior end, in the cloacal region (Fox, 1970). A detailed description of the histology of the gut epithelium is given by Bonneville and Weinstock (1970).

The development of the pronephric kidney in *Xenopus* has been described by Fox (1963) who compared it with that of other amphibians to see if it could be regarded as more primitive in *Xenopus*. He found that *Xenopus* has three pairs of pronephric tubules only, as in other Anura, and that these arise in segments 7 to 9, becoming functional two days after hatching. Fox and Hamilton (1964b) studied experimentally the mode of development of the pronephric duct. They joined anterior and posterior halves of neurulae, in some cases with one half rotated 180° to see whether the backward-growing duct from the anterior half would then deviate, as Holtfreter (1943) had shown it to do in other amphibians. In other cases Fox and Hamilton joined diploid halves to haploid halves, so that the extent to which the posterior half contributed cells to the duct could be recognized by chromosome counts. They found that in *Xenopus*, unlike other amphibians, posterior parts of the duct grow independently in the posterior half and then join up with the anterior portion at a level much further cranial than in other species. They suggest that the posterior parts of the duct arise either from an extended diverticulum of the cloaca, or by differentiation *in situ* in the posterior half.

Recently the ultrastructure of the kidney has been compared between haploid and diploid *Xenopus* larvae, by Fox and Hamilton (1971). They found that differentiation of the pronephros was retarded in haploids and that their cells contained fewer mitochondria than those of diploids.

Genital System.   There has been relatively little recent work on the development of the genital system in *Xenopus*, other than the tests of effects of parabiosis and of hormones on larvae, discussed in Chapter 3. Michalowski (1959) described an anomalous case of hermaphroditism in which there was an ovary on one side and a testis on the other, but the aetiology of such anomalies has not been investigated. Organ culture of gonads shows that they are capable of self-differentiation if initial signs of development into either ovary or testis are already present before culture (Foote & Foote, 1960). Their differentiation can, however, be influenced by hormone treatment *in vitro*. Chorionic gonadotrophins stimulate hyperplasia of the rete apparatus in both sexes: androgens retard ovarian growth, but do not cause reversal to a testis: oestradiol has no effect on the testis. In view of the earlier evidence which has suggested that gonadal sex in lower vertebrates is readily reversed by hormone treatment, these findings by Foote and Foote raise several new problems. Clearly there is a need for much more work on the effective hormone dosage for reversal of sex at different stages of gonadal development in amphibians, both in organ culture and *in vivo* (cf. Chapter 3, section 4).

4. Tissue Interactions Studied in Organs of *Xenopus*

(a) The Eye

The phenomenon of 'neural induction' which was discussed in Chapter 8 really embraces development of the eye too, for this organ is an integral part of the central nervous system. The eye, like the brain and spinal cord, requires an inducing stimulus from the underlying mesoderm to initiate its development. But besides this there are secondary interactions which control the development of the component parts of the eye. The fact that the eye vesicle induces the ectoderm with which it comes into contact to form a lens, was discovered by Spemann (1903), long before he demonstrated the induction of the neural plate by dorsal lip mesoderm. Spemann himself carried out some of the first experiments in which an eye vesicle was transplanted into another site in the newt embryo and was found to induce a lens. It was this work, in fact, which led him to envisage the possibility of induction occurring between other embryonic tissues.

In more recent experiments on *Xenopus*, somewhat conflicting evidence has been obtained as to the ability of the eye vesicle to induce a lens. Brahma (1959) showed that an eye cup transplanted ventrally at the tailbud stage induced a lens successfully in ventral ectoderm. But he found that ectoderm removed from over a normally sited eye at this stage was not yet capable of self-differentiation into a lens. So apparently a lengthy duration of contact between eye vesicle and ectoderm was necessary, for lens induction to be completed. On the other hand, Babcock (1963) found that removal of the eye primordium did not prevent a lens from forming in the head region, even if the removal was as early as the neurula stage, before the eye vesicle had formed. So it seems probable that other factors from the head region are also involved in induction of the lens in *Xenopus*. The 'free lenses' which formed *in situ* without participation of the eye cup in Babcock's experiments developed abnormally, however. Instead of invagination, there was an aggregation of cells within the lens placode, and there was not the same increase in RNA as in normal lens cells.

(b) The Limb

The interactions between mesoderm and ectoderm in the developing limb have been analysed more thoroughly in *Xenopus* than in most other amphibians. Tschumi (1957) traced the fates of different regions of the limb bud of *Xenopus*, by marking the cells with carmine grains. He showed that most of the early elongation of the bud was due to addition of cells at the distal end, rather than to intercalary growth. Tschumi's marking experiments are illustrated in Figure 9.5. He also tested the effects of the apical ectoderm of the limb bud on the development of the mesoderm (Tschumi, 1956), since it is well known that in the chick, an apical ectodermal ridge is responsible for controlling the orientation of the limb and for inducing the formation of digits. Tschumi pointed out that a similar ridge is present in *Xenopus*, and he found that its removal prevented development from going to completion. He also carried out rotations of the apical ectodermal cap, and found that this altered the orientation of the marginal vein and hence of the digits

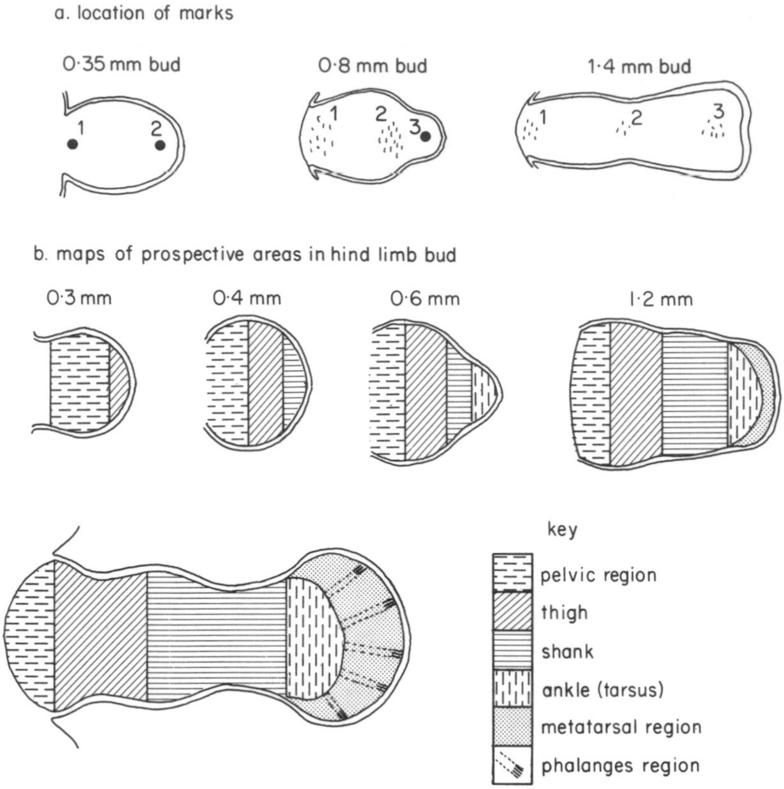
a. location of marks
0·35 mm bud
0·8 mm bud
1·4 mm bud
1
2
3
b. maps of prospective areas in hind limb bud
0·3 mm
0·4 mm
0·6 mm
1·2 mm
key
pelvic region
thigh
shank
ankle (tarsus)
metatarsal region
phalanges region

Figure 9.5. Results of marking experiments on cells of the hind limb bud in *Xenopus*. Redrawn from Tschumi 1957. (a) top, row: movements of marked cells as bud grows. (b) bottom two rows: maps of the areas that will form each region of the limb, deduced from marking experiments

(Figure 9.6). This experimental work still stands alone: others have not carried it further, perhaps because the chick embryo is more convenient than the amphibian larva for operations on limb-buds. Recently however, Tarin and Sturdee (1971) have confirmed the existence of the apical ectodermal ridge in *Xenopus*. They have also described its histochemical characteristics (Tarin & Sturdee, 1973), but these do not give any clues as to the mechanism of its action on the mesoderm.

5. Concluding Remarks

Because of its modifications for a fully aquatic habit, *Xenopus* has not been used as a 'type' from which amphibian morphogenesis was first described. Hence we do not find a large body of descriptive literature on organogenesis

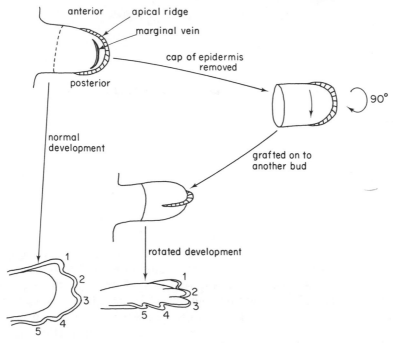

Figure 9.6. Results of rotating apical cap of limb bud ectoderm in *Xenopus*. Modified from Tschumi, 1956. The marginal vein is displaced and the digits reorientated to conform with the rotation

in this species. We do find, however, that since it became realized that *Xenopus* is particularly convenient for laboratory work, interest has been aroused in those of its organs and morphogenetic processes which are unusual because of its aquatic life. Hence attention has been paid specially to the lateral line organs and to other features of the anatomy of the adult which show differences from other amphibians (cf. also Chapter 2). In the larva, the specialized gut and the readily-visible melanophores have attracted investigators. Comparative and experimental embryologists have also wanted to see in what ways the developmental mechanisms in organogenesis differ in *Xenopus* from those of other species. As we have seen here, there are similarities in general, but differences of detail.

In the next and last main chapter, we shall consider how some of the organ systems interact in later development and at metamorphosis.

10

The Control of Later Events in Development

The last chapter included a few examples of how the component parts of some organs interact with each other as they develop. We shall now go on to discuss some ways in which different organs interact during their later development in the larva, at metamorphosis and in the adult frog. Some well-known examples of organ interactions which have been studied further in *Xenopus* recently are: (1) the relationships between the central nervous system and end-organs, (2) factors controlling regeneration, (3) the influences of endocrine glands on each other and on other tissues during metamorphosis, and (4) the development of the immune system. It is easiest to consider these topics under separate headings, though some of them overlap considerably.

1. Interactions between the Central Nervous System and End-Organs

It is a familiar fact that the development and maintenance of both sensory and effector organs in the body depend on an adequate nerve supply. Many instances of this are seen at later stages in life, when an organ begins to atrophy if the nerves supplying it are damaged. In the embryo, the formation and growth of these end-organs are influenced in many ways by the nervous system, and there is also a reciprocal action of the end-organs on the maturation of neurons in the central nervous system. Much of the experimental work on these interrelationships between the nervous system and end-organs has been carried out on amphibian embryos, and the most recent findings have been made on *Xenopus*. Particular attention has been paid to the relationships of the eye and the limb to their central connections in the brain and the spinal cord.

(a) Interactions between the Brain and the Eye.

Since the eye originates as an outgrowth from the forebrain, it is perhaps not surprising that the development of these two organs is found to be linked.

To begin with, they control each other's growth rate. This was shown in early experiments by Balinsky (1958) who exchanged brain and eye tissue between embryos of different genera of Anura, and found that the ratio of brain size to eye size typical of each genus was maintained, even in the transplants. One particular region of the brain which is later influenced most by the eye is the optic tectum, to which sensory fibres of the optic nerve connect. McMurray (1954) showed that destruction of the optic nerve on one side in *Xenopus* caused a 50 % reduction in size of the optic tectum on the contralateral side of the brain. The great majority of optic nerve fibres in *Xenopus* cross over at the optic chiasma (see Figure 10.1) to enter the tectum of the opposite side: so

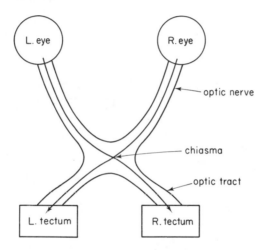

Figure 10.1. Diagram of the optic chiasma and the main nerve pathway from retina to optic tectum in *Xenopus* (cf. Figure 2.9(b))

evidently in the absence of these fibres, the optic tectum was not receiving some stimulus that was necessary for its full development. This supposition was confirmed by further experiments in which McMurray simply crushed the nerve on one side. This caused a temporary hypoplasia in the optic tectum, but it was restored to normal as soon as the nerve fibres regenerated through the crushed region.

A number of elegant mapping experiments, using microelectrodes to record nerve impulses at discrete points on the tectum, have shown that sensory impulses from each region of the retina of the eye project onto a precise area of the optic tectum. Gaze and Jacobson and their associates have devised such maps of the retinotectal projections in *Xenopus* (see Figure 10.2). They have also studied the effects of severing the optic nerve and rotating the eye 180° before rejoining it to the nerve stump—then allowing

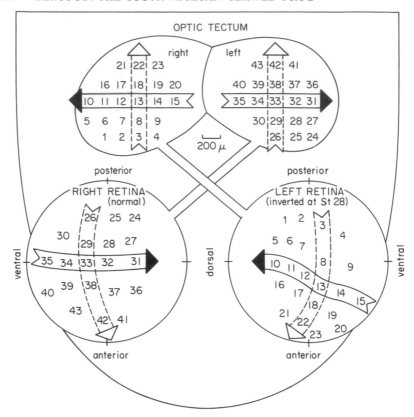

Figure 10.2. Diagram of the retinotectal projections from right and left eyes in *Xenopus*, after Jacobson 1968. The left eye was rotated at a stage before the critical time when the retinal axes are fixed, and so its projections onto the right tectum are normal

the fibres to regenerate. Experiments of this kind were first performed by Sperry (see Sperry, 1965) on other amphibians, and he deduced from his results that the optic fibres made accurate point-to-point reconnections with their proper area of the optic tectum. This seemed to him the best way to explain why most of his animals showed reversed visual responses: i.e. when an object was placed above them, they behaved as if they saw it below, and vice versa. So Sperry promoted the 'neuronal specificity theory', according to which each neuron must have some chemical specificity which makes it connect only with certain other neurons during development, and these chemical specificities govern the development of all interneuronal connections in the embryo. However, the experiments on *Xenopus* by Gaze, Jacobson and their collaborators (reviewed in Gaze's book, 1970) led them to different conclusions. Jacobson (1965) followed by electrophysiological mapping the reconnections that were made between retinal cells and points on the optic tectum after the eye had been rotated. He found that if the rotation was done before stage 30 (late tailbud stage), connections were made as if the eye had not been rotated at all. Rotation of the eye after stage 30, however, led to a reversal of the retinotectal map, and reversed behaviour too. This last result might seem to agree with Sperry's findings: however, when the course of regeneration of the optic fibres was followed in detail, it was found that their growth was very disorganized at first and gave no evidence of any point-to-point specificity such as Sperry first envisaged. There was only a gradual acquisition of an orderly pattern, with some breaking of original connections and making of new ones during this process.

Another ingenious experiment of Sperry's, in *Rana* larvae, had been to make a sphere by folding together the edges of one nasal or one temporal half of an eye, then to rejoin this spherical half-eye to the optic nerve. He claimed that the optic fibres then grew only into their appropriate, nasal or temporal half of the optic tectum, leaving the other half uncolonized. However, Gaze, Jacobson and Székely (1965) obtained different results in *Xenopus*. They fused two nasal or two temporal halves to form 'compound' eyes, and from each type they found that the optic fibres spread over the whole of the optic tectum, showing no preference for the half appropriate for nasal or temporal retina (cf. Figure 10.3). It could have been argued that

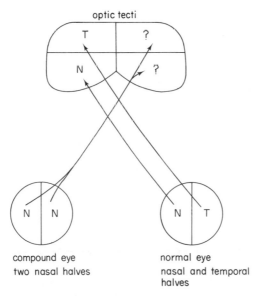

optic tecti

compound eye
two nasal halves

normal eye
nasal and temporal
halves

Figure 10.3. Diagram to show the projections from a double-nasal compound eye to the tectum: simplified from Gaze, Jacobson and Székely, 1965

perhaps the tectum itself was abnormal because of the abnormal eye connecting with it, and that only half of its areas had developed. However, Gaze, Jacobson and Székely (1965) noted that double-nasal eyes sent some fibres to the optic tectum of the ipsilateral side, which normally only receives a few temporal fibres from its own side. In later work (Straznicky, Gaze & Keating, 1971) they uncrossed the optic nerves after some fibre growth had begun, so that the fibres from the compound eye were now supplying the tectum which had been connected with the normal, control eye previously (cf. Figure 10.4). In this case too, the whole of the tectum was colonized by fibres from the compound eye, with no evidence of specific connections being made to only some areas. So it seemed no longer possible to apply any 'neuronal specificity' theory to retinotectal connections. There is still some controversy about this, however, and recently Straznicky (1973) has opened up again the possibility that there is some topographical specificity in the connections, for he found that if parts of the tectum are removed, certain areas of the retinae appear to be blind and have evidently not been able to make connections with the tectum.

Straznicky and Gaze (1971, 1972) carried out in addition some autoradiographic studies of the sequence of neuronal development in both retina and optic tectum in *Xenopus*. They found two most surprising things: first, that at stage 30 (when the regions of the retina seem, from the results of rotation experiments, to become specified) only a very small group of sensory cells in the centre of the retina are mature. For a long time after Stage 30, cells are still being added at the periphery of the retina. Labelling with tritiated thymidine shows that concentric layers of cells, gradually more peripheral, are added until at least stage 45. Their second unexpected finding was that in the optic tectum, there is not any concentric pattern of neuron maturation as in the retina, but the sequence of maturation is from rostroventral to caudomedial (Figure 10.5). So it does not seem possible to suggest any point-to-point relationship between retinal and rectal cells during their development. In a recent discussion Gaze and Keating (1972) suggest that there is a broader, more flexible pattern of nerve connections, which they term 'systems matching'. It will be difficult to design experiments to demonstrate such a mechanism for retinotectal connections in embryos, however.

tectal

(b) Development of Spinal Sensory and Motor Connections and their Interactions with the Limbs

Hughes (1957, 1959) has established some interesting points about the normal development of sensory and motor cells in the nervous system of *Xenopus* (see his book, 1968, for review and comparison with other vertebrates). He showed that the sensory elements in the spinal cord first appear as a series of dorsal, extramedullary neurons which migrate out from the spinal cord and have dendrites in contact with the myocommata between the

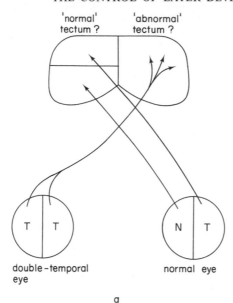

Figure 10.4. Experiment in which optic nerves were uncrossed, after one tectum had been receiving fibres from a compound eye. (a) the original connections: (b) after uncrossing the nerves, fibres from both eyes still connect with all parts of the optic tecta. Simplified from Straznicky, Gaze and Keating, 1971

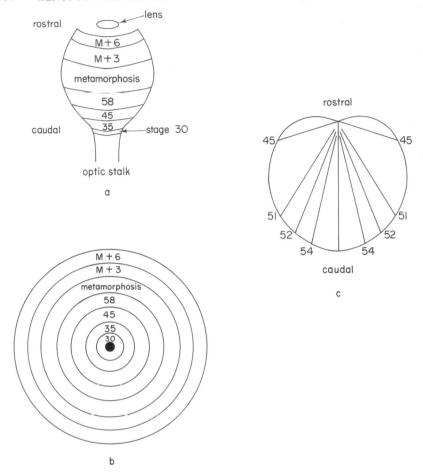

Figure 10.5. Diagrams to show the sequences of maturation of cells in the retina and the optic tectum in *Xenopus*. Modified from Straznicky and Gaze, 1971, 1972. (a) Retina, viewed as if in horizontal section; (b) Retina, viewed through pupil; (c) Tectum, viewed from dorsal aspect. Figures represent stage numbers according to Nieuwkoop and Faber's Normal Table, 1956

myotomes (Hughes, 1957). These cells are not seen in other species, but the large, dorso-lateral Rohon-Beard cells which appear later within the cord and persist until the dorsal root ganglia make connections with the cord, are found in other amphibians (Figure 5.12, p. 117). Hughes (1959) also showed some peculiarities in the motor elements in *Xenopus*. Each early motor axon grows caudalwards for some $1\frac{1}{2}$ segments before emerging in a ventral nerve root. As it passes caudalwards it may anastomose in the cord with other axons from adjacent segments: this led some observers to believe that there were precocious spinal tracts in *Xenopus*. There are segmental motor roots by the mid-larval stage, and swimming movements then become controlled by post-otic centres in the brain (Sims, 1962).

It has been known since the classic work of Detwiler (1936) that the morphology of the spinal cord is influenced by the positions of the developing limbs in vertebrates. Detwiler showed in the salamander *Ambystoma* that the motor horns of grey matter in the spinal cord became enlarged in those segments that were opposite the limb buds, and that removal or transplantation of the limb buds led to enlarged motor horns at whatever levels the limbs now were. Recently a quantitative study of the relationships between the numbers of neurons developing and the presence or absence of a limb bud at various stages has been carried out in *Xenopus* larvae by Hughes and his associates.

In earlier work, Hughes and Tschumi (1958) and Hughes and Lewis (1961) showed that removal of the limb buds, or suppression of their development by treatment with a nitrogen mustard, led to a reduction in the numbers not only of motor horn cells but also of sensory ganglion cells in the limb segments of the spinal cord. This was the first time that sensory elements had been found to be affected as well as motor ones. After removal of a limb on one side, there was first a chromatolysis of motor horn neuroblasts and then a reduced number of neurons later, both in the motor horns and in the corresponding sensory ganglia. The same effect was produced by transplanting the limb bud ventrally so that the segmental nerves could not reach it (Hughes & Tschumi, 1960) or by introducing blocks of mica in the path of the nerves. In all these cases, the normal number of neuroblasts formed in the spinal cord, but more degenerated than normally, so that there were eventually fewer mature neurons. Prestige (1965) made a detailed study of the extent to which neuroblasts degenerate in *Xenopus*, in both normal and experimental conditions. He found that as many as two-thirds of the neuroblasts in the sacral ganglia (not at the limb level) normally degenerate between stages 53 (late larva, with limb buds at 'paddle' stage) and 59 (mid-metamorphosis). In the trunk region, some 4,000 neuroblasts degenerate during this time. So evidently the limb ablation experiments simply enhance this normal process. In birds too, Hamburger and Levi-Montalcini (1949) showed that there is normally a large-scale death of neuroblasts during the development of the central nervous system. In *Xenopus*, Prestige (1967) went on to show that the response of the ventral horn cells to limb amputation varied with their state of maturity. They underwent chromatolysis more quickly if the limb was removed at their later stages of maturation.

There is not much evidence about the reciprocal influence of the nervous system on the development of the limbs in *Xenopus*. Most of our evidence that nerves affect limb development comes from organ culture experiments, carried out with chick limb buds (e.g. Bradley, 1970) and from experiments on urodeles. In the absence of nerves, cultured limbs are stunted compared with normal limbs and their joints tend to fuse. If the neural plate is removed from a urodele embryo and it is then joined to a normal embryo to enable it to survive (Piatt, 1942), its limbs will develop without innervation ('aneurogenic' limbs). These limbs, too, are stunted. No equivalent experiments have been done on *Xenopus*, but it is well known that limb regeneration in this animal depends on an adequate nerve supply (see section 2 below). Little is known about the early control of limb function in *Xenopus*, but Hughes and Prestige (1967) noted that in the hindlimb, a stretching or 'flare' movement occurred before any sensory function could be demonstrated. This suggests that motor nerves become functional before sensory ones: a phenomenon that is generally true in other vertebrate embryos studied.

2. Regenerative Processes and their Control

Under the heading of 'regeneration' one usually includes only those replacements of organs and tissues that occur after the embryonic period is over, although we have met several examples of the embryo being able to replace lost parts. Such adaptive processes are not so surprising in the embryo as they seem later on, when one would not think it possible for cells of fully-formed tissues and organs to start differentiation again. Nevertheless they can in some circumstances do this, and there is a remarkable variety of ways in which regeneration occurs in different organs and in different animals. Amphibians have much greater powers of regeneration than birds or mammals, and can replace limbs, tail, lens and many internal organs. Most experimental work has been carried out on the appendages and the lens, and there are some special features about the regeneration of these organs in *Xenopus* that we may now consider briefly. Full accounts of regeneration in amphibians may be found in Balinsky (1970) and Needham (1942).

(a) Regeneration of the Limb

The pattern of regeneration of the limb is similar in all Amphibia. First, nearby epidermal cells move to close over the cut surface, then a process of demolition removes dead cells into the blood stream. Next, the mesoderm undergoes a remarkable process of 'dedifferentiation', forming rounded cells which accumulate in a mass at the tip of the limb stump, under the epidermis, forming the 'blastema' (Figure 10.6). This blastema then gradually redifferentiates into tissues that reconstitute the lost parts of the limb. Sometimes blastema cells that have been traced by radioactive labelling and are known to have been derived from one tissue—for example, cartilage— redifferentiate into some other tissue such as muscle: so they show considerable versatility, just as if they had re-acquired the properties of embryonic cells. Burgess (1967) tried to see if their nuclei had become 'embryonic', by

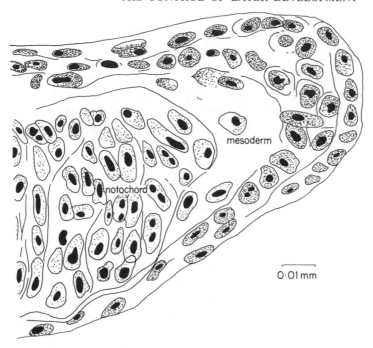

Figure 10.6. Camera lucida drawing of sagittal section of regeneration blastema of the tail in *Xenopus*, after Dettelbach, 1952. Mesoderm and notochord are indicated. Blastema cells appear just below the epidermis

transplanting these into enucleated eggs to see if they would allow normal development, like blastula or gastrula nuclei (cf. Chapter 6). But she found that the blastema nuclei did not work in this way: so evidently one cannot regard them as having re-acquired all the potentialities of early embryonic nuclei.

The regenerative power of the limb in Anura is more limited than in Urodela and declines sharply after metamorphosis. *Xenopus* appears to regenerate limbs better than other Anura, however. This could be interpreted as evidence either of its affinity to the Urodela (cf. Chapter 2) or of its primitive characteristics. If the limbs of adult *Xenopus* are amputated at various levels (Gallien & Beetschen, 1951), regeneration is not always complete, however: it is more successful at distal than at proximal levels. Both fore and hindlimbs can regenerate digits perfectly, but only a short conical segment is formed if the forearm or lower leg is amputated, and there is no regeneration at all from upper arm or upper leg levels. In metamorphosing larvae these upper regions of limbs can, however, be regenerated if the amputation is performed before stage 51 (when they are still only simple buds).

From this stage on there is a decline in the regenerative ability of the limbs. Even so, their regenerative powers in *Xenopus* greatly surpass those of other Anura, which cannot replace any limb parts at all in adult stages.

It has been shown by Singer (1952) that the peripheral nerve supply to a limb stump influences its rate of regeneration. He has for some time collected evidence suggesting that there is a quantitative relationship between the total axonal volume supplying the limb, and its rate and success of regeneration. If other nerves are diverted to a limb stump in addition to its own supply, regeneration is improved, or a supernumerary limb may even form, while conversely if the limb stump is denervated, regeneration is inhibited. On Singer's theory, a possible reason for the superior regenerative powers of limbs in *Xenopus* is that this animal has smaller, more numerous nerve fibres than other Anura, adding up to a larger total volume of nerve axons than in the fewer, coarser nerve fibres of other species.

(b) Regeneration of the Tail

Since the tail is present throughout larval life, much more work has been done on its regeneration than on the limbs, which do not appear till near metamorphosis. The course of tail regeneration has been described fully for *Xenopus*, both in morphological terms (Dettelbach, 1952) and in biochemical terms by Lehmann, Weber and their collaborators (see Lehmann, 1957 and Weber, 1967 for reviews). In fact not only normal growth and regeneration, but also the regression of the tail at metamorphosis has been followed by these workers. This single organ is specially convenient for studies of growth, repair and regression within a short span of time. For example, the roles of certain enzymes in all of these processes could be followed in the tail. One class of enzymes whose role was problematical in the 1950's were the cathepsins, which appeared to have synthetic as well as proteolytic functions (Fruton *et al.*, 1953). From measurements of catheptic activity in tail stumps and regenerates in *Xenopus*, related to the trends in total nitrogen and in free amino acid concentrations (Jensen, Lehmann & Weber, 1956; Deuchar, Weber & Lehmann, 1957) it became clear that high catheptic activity was associated with tissue breakdown, which preceded the formation of the blastema in the tail stump. Weber (1957) went on to show that there was a dramatic rise in catheptic activity in the tail at metamorphosis, when it began to regress as a result of large scale demolition processes (see section 3 (c) below). So these results emphasized the proteolytic, rather than synthetic role of cathepsins.

(c) Regenerative Processes in the Nervous System

Amphibians are capable of regenerating parts of the brain and spinal cord, even at quite late stages in larval life in Anura and at adult stages in Urodela. A transection of the spinal cord will cause outgrowth of new nerve fibres in place of the damaged ones, until eventually a fairly normal cord is reconstituted and normal function returns. In *Xenopus*, Sims (1962) studied the rate of regeneration of fibres of the spinal cord after transection between the 5th and 6th segments, at stage 56 (near metamorphosis). He found that nerve axons crossed the gap within seven days.

Recent work by Hauser (1965, 1969) has shown an interesting relationship between regeneration in the central nervous system and that of the tail. Transection of the brain just behind the midbrain region causes only a small and incomplete tail to regenerate in a larva whose tail has been amputated. If the brain ventricles are occluded, or if the subcommissural organ which lies in the roof of the midbrain is damaged, tail regeneration is completely inhibited. Earler, Skowron, Jordan and Rogulski (1956) had reported that elimination of the choroid plexus (a procedure which may also have damaged the subcommissural organ) retarded tail regeneration. The subcommissural organ (Figure 10.7) is found in fish as well as Amphibia and is thought to be neurosecretory (Olsson, 1955). It sends long extensions, known as Reissner's fibres, through the neural canal right to the tip of the spinal cord. Hauser (1972) has now shown that the extent of regeneration of the tail *in situ* depends on the condition of these Reissner's fibres, which are also capable of regeneration after damage. If the fibres are restored by the time a tail blastema has formed, a normal tail will regenerate. Despite these findings, it cannot be said that tail regeneration is entirely depe₁dent on Reissner's fibres, however, since tail stumps isolated *in vitro* h₁ve been shown to regenerate quite well (Hauser & Lehmann, 1962).

A part of the central nervous system which is of special interest to students of regeneration is the eye. Parts of the retina can regenerate, if the eye cup is damaged at early stages before its differentiation is complete. The lens can be regenerated at much later stages, even up to the adult, in Urodela and in *Xenopus*, but not in most other Anura. In urodeles the lens is usually regenerated from the iris or the retina, in the dorsal quadrant of the eye (Stone, 1954). In *Xenopus* too, lens regeneration most frequently takes place from the dorsal iris (Overton & Freeman, 1960) but it may also occur from the cornea (Freeman, 1963). Campbell (1963) compared the frequencies of lens regeneration from different parts of the eye in *Xenopus* and found that in larvae, 12 % of lenses were formed from the cornea, 12 % from the iris and 56 % from the neural retina near the ciliary body. But in adults, regeneration was always from the dorsal iris, indicating that there was by then less versatility in the different parts of the eye. The regeneration also took longer in adults: about 6 months instead of 14 days as in the larva.

Corneal lenses do not form from a hollow vesicle as in embryonic development, but arise from a solid ball of cells which have delaminated from the inner surface of the cornea. This is reminiscent of the mode of formation of 'free lenses' (cf. Chapter 9). Overton (1963) found, however,

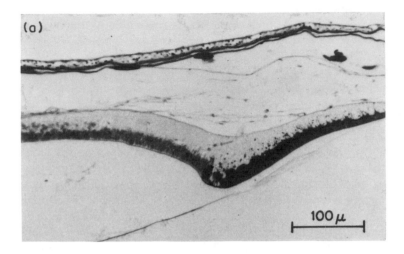

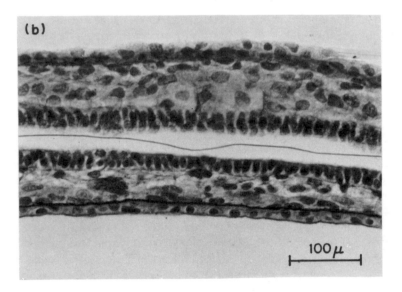

Figure 10.7. Longitudinal sectional views of the subcommissural organ and Reissner's fibres in *Xenopus* larvae: originals from Hauser. (a) Roof of midbrain, with dark-staining columnar cells of the subcommisural organ showing prominently, and a fibrous extension from them into the brain ventricle below: this is the beginning of the Reissner's fibres. (b) Part of the central canal of the spinal cord, with a Reissner's fibre running in it.

that the ultrastructural changes in these regenerating lenses were closely similar to those in the embryonic lens. So evidently the cytological events in lens-formation are less variable than histological or morphological ones. The same seems to apply to biochemical events such as the appearance of new proteins. Immunological work on other species of Amphibia and on birds has shown that the same sequence of α-, β-, and γ-crystallins appears in regenerated lenses as in embryonic development. Campbell (1965) showed that in *Xenopus*, although there are no lens antigens in the cornea of unoperated eyes, some lens antigens begin to appear by 24 hours after lentectomy, in those inner corneal cells that are due to form the new lens.

3. Interrelationships Between Organs at Metamorphosis

The remarkable phenomenon of metamorphosis in Amphibia involves large-scale morphological alterations such as resorption of the tail and development of limbs, as well as smaller-scale though no less important changes such as restructuring of the skin and modification of the absorptive surfaces of the gut. In many Anura, another major change is the development of lungs, but in *Xenopus*, as we have seen already (Chapters 2 and 5), lungs are already present in the larva. Some of the metabolic changes typical of metamorphosis in Anura also do not occur with quite the same abruptness in *Xenopus*, since it remains aquatic as an adult and is, for instance, able to continue excreting a high proportion of ammonia rather than changing over completely to urea excretion.

A full description of morphological events at metamorphosis in *Xenopus* has been given by Gasche (1944), and essential features are mentioned in the Normal Table by Nieuwkoop and Faber (1956). The changes in external appearance at this time were summarized in Figure 5.13.

Preparation for metamorphosis are set in train by stimuli from the pituitary and thyroid glands, which interact during their functional maturation in larval stages. We will start, therefore, with a brief consideration of some of the findings relating to endocrine function in premetamorphic larvae of *Xenopus*.

(a) Endocrine interrelationships during metamorphosis

The thyroid gland develops in *Xenopus*, as in all other vertebrates, as an outgrowth from the floor of the pharynx. The pituitary develops, again typically of vertebrates, partly from an ectodermal upgrowth in the roof of the buccal cavity—this part forms the anterior pituitary, or adenohypophysis —and partly from a downgrowth from the floor of the forebrain which forms the posterior pituitary, or neurohypophysis. In *Xenopus* as in all other vertebrates studied, the thyroid gland depends for its normal development and functioning on factors reaching it in the blood stream from the pituitary gland and also on the hypothalamus, a specialized region in the floor of the forebrain which has neuroscretory cells and is an important coordinator of endocrine function. Goos (1969) distinguished two cell types responsible for the control of thyroid function in *Xenopus*: one type in the pituitary and the other type in the hypothalamus. These cells secrete, respectively, a thyroid-stimulating hormone (TSH) and a thyroid-releasing factor (TRF). There are also some mucoid cells in the pituitary whose activity seems to be linked with that of the thyroid (Saxèn, 1960). Removal of the pituitary from an amphibian tadpole causes the thyroid to remain small and to fail to secrete thyroxin. As a result, the tadpoles fail to metamorphose, becoming what is known as 'neotenic' if their genital system matures while they are still in a larval state. Neotenic *Xenopus* tadpoles can also be produced in the laboratory by insufficient lighting (Toivönen, 1952) and these show a decrease in the number of basophil cells in the pituitary, besides having small thyroids, deficient in iodine. Kerr (1965) observed that basophil and acidophil cells were present in the anterior pituitaries of normal *Xenopus* larvae by stage 47 (full-grown larva but with no limb buds yet). He was not able to show directly whether the activity of these cells was linked with that of thyroid, but he noted that both types of cell increased in number and size just before metamorphosis, then shrank, as if they had discharged their contents, when metamorphosis began. Kerr's main interest was in the gonadotrophin-secreting cells of the pituitary, which are widely distributed after metamorphosis (cf. Figure 3.4, p. 69).

Saxèn and collaborators (1957a, b) and Coleman, Evennett and Dodd (1968) have followed in some detail the histological changes in the thyroid and its iodine uptake, in relation to changes in the pituitary, during larval, metamorphosing and young adult stages in *Xenopus*. They found that both glands underwent increased hormone synthesis and storage just before metamorphosis, reaching a maximum at the stage when the forelegs emerge, which is regarded as the 'metamorphic climax'. More recently Neuenschwander (1972) has used electron microscopy and labelling with radioactive iodine (125-I) to show that the thyroid cells both increase in size during the metamorphic period and increase their uptake of iodine. According to Saxèn and coworkers (*loc. cit.*), protein stores become depleted in both pituitary and thyroid glands during the metamorphic period, and there is a net increase of protein-bound iodine in the tissues at the expense of the thyroid.

It has been questioned whether the less complete metamorphosis shown in *Xenopus* as compared with other Anura, with retention of lateral line organs, ammonia excretion and an aquatic habit, could be due to inadequate thyroid function. Alternatively the tissues might respond inadequately to thyroxine, like those of axolotls (Deuchar, 1957). Newth (pers. comm. 1971) found that it was not possible to alter any of the features of metamorphosis in *Xenopus* by increased doses of thyroxine, however. It is possible anyway that such doses would have depressed the function of the animal's own thyroid, since Fox and Turner (1967) showed that the size of the thyroid in *Xenopus* tadpoles was inversely related to the concentration of thyroxine added to the medium. Conversely, addition of thyroxine antagonists such as phenylthiourea caused enlargement of the thyroid in an attempt to compensate: this was insufficient to allow metamorphosis to occur, however.

There is a pair of apparently endocrine organs called the 'ultimobranchial bodies', derived from the 6th pharyngeal pouch on each side, which come to lie within the tissue of the thyroid. Saxèn and Toivönen (1955) found that these bodies do not take up iodine, nor do they develop follicles like the thyroid. So the function of these organs in *Xenopus* remains unknown, though there is evidence that in birds and mammals, they secrete a hormone called 'calcitonin' which regulates the deposition of calcium in the body. Another problematical organ in *Xenopus* is the pineal gland in the roof of the midbrain. Haffner (1951) described its development in detail and noted that it reaches its maximum size at metamorphosis and then degenerates, acquiring fatty deposits in place of its cells. Bagnara (1963) believed the pineal to be a source of the hormone melatonin, which helps to control the movements of pigment in the melanophores (cf. Chapter 3). Since Baker and Hoff (1971) found that over 90 % of the melatonin extractable from *Xenopus* larvae is contained in the eyes, this possible function of the pineal has been thrown into doubt. There seems some evidence that it may be a light-receptor, like the pineal of reptiles (Stebbins & Eakin, 1958). In late larval life the pineal cells become transparent and a layer of pigment cells separates them from the midbrain below (Haffner, *loc. cit.*). Ultrastructural studies show that there are photo-receptor cells present as well as secretory ones (Charlton, 1968). Bogenschütz (1965) suggested that the pineal is responsible for the ability of eyeless larvae to respond to changes of background colour—a response which is very slow in eyeless adults whose pineal has begun to degenerate (Vilter, 1946; cf. Chapter 3). But it is difficult to envisage how an organ which is situated dorsally could detect background changes which lie below the larva. Obviously further investigation of the function of the pineal in *Xenopus* is much needed. It is still unknown whether it plays any role at all in metamorphosis.

(b) Responses of Tissues to Thyroxine at Metamorphosis

One of the most enigmatic features about metamorphosis is that the several organs and tissues of the larva respond in such different ways to the one hormone, thyroxine which initiates the events. That this hormone, and no other systemic factor, is the essential stimulus, has been shown by treating various tissues *in vitro* with thyroxine. Thus, Weber (1962) showed that tail-tips of *Xenopus*, which can be maintained in culture for long periods (Hauser & Lehmann, 1962) will regress if thyroxine is added to the medium.

Shaffer (1963) obtained similar results using the thyroxine derivative, tri-iodothyronine. Since the tail responded to these agents *in vitro*, it clearly did not require other systemic factors as well to make it regress.

There are some very subtle discriminations in the effects of thyroxine on different tissues and regions of the body in amphibian larvae. For instance, Gross (1964) showed that in response to implanted crystals of thyroxine, skin of the trunk in the amphibian larva would start to lay down extra collagen, even if this was near the borderline between trunk and tail. Just the other side of this borderline, an implanted crystal under the tail skin caused it to start the very opposite process, of collagen breakdown as it would during regression at metamorphosis. Another subtle local difference is shown in the response of brain cells to thyroxine. Weiss and Rosetti (1957) showed that implants of rat thyroid gland into the midbrain of *Xenopus* tadpoles caused most of the cells to start mitosis, but not the giant Mauthner's cells in the hindbrain, which normally degenerate at metamorphosis: these, on the contrary, began to degenerate in response to the thyroid implant. The degeneration is preceded by the appearance of lysosomes in the cytoplasm (Moulton, Jurand & Fox, 1968). The giant neurons have axons extending the full length of the spinal cord and are used in rapid darting movements by which the larva escapes predators.

Rather less detail is known about events in the limbs of *Xenopus*, which start to develop in response to thyroxine. Fox and Irving (1950) showed that thyroxine causes alignment of cartilage cells in the femur, and osteogenic processes prior to the formation of a marrow cavity are accelerated. Later changes in the skeleton appear to be independent of thyroxine.

We shall now go on to consider some of the changes in metabolism that are an essential part of metamorphosis in *Xenopus*.

(c) Biochemical Events during Metamorphosis

(i) Breakdown Processes in the Tail. An organ in which most dramatic biochemical activity occurs at metamorphosis is the tail, as it regresses. It has already been noted (p. 190) that one major event is a steep rise in catheptic activity (Weber, 1957). The affinity of the enzyme for substrate also increases during this period, as evidenced by a change in its Michaelis constant (Weber, 1965): this may be important in enabling the rate of tissue destruction to be maintained despite a decrease in total protein and an accumulation of breakdown products. Benz (1957) showed that the highest catheptic activity was at the tail tip, which is the starting point for regression. Many other catabolic enzymes accumulate in the regressing tail tissue: Eeckhout (1969) located acid hydrolases in the connective tissue, including catalase and acid phosphatases. Von Hahn (1958) found that catalase activity was also highest at the tail tip. Weber and Niehus (1961) observed a steep rise in acid phosphatase activity at the onset of tail resorption, paralleling the rise in

catheptic activity. Recently Perriard and Weber (1971) have observed increases in free amino acids too, which indicate protein breakdown, just before metamorphosis begins.

Ultrastructural studies (e.g. Weber, 1964) show the presence of lysosomes in those cells that are autolysing in the tail. Vitamin A is an activator of lysosomal enzymes in skin, and it has been found also to induce some metamorphic changes in tail tissue (Weissmann, 1961). The details of regression in the muscles, with folding of myofibrils and a gradual disappearance of the I-bands and mitochondria, have been described by Weber (*loc. cit.*).

Enzymes concerned in carbohydrate metabolism tend to decrease in activity in the tail during metamorphosis: mitochondrial ATP-ase activity, for instance, decreases (Niehus & Weber, 1961). An interesting general point about these destructive processes is that they are evidently determined genetically. Any treatment which blocks the transcription of DNA will therefore delay the process of resorption of the tail: for instance this resorption is inhibited by Actinomycin D in both *Xenopus* (Weber, 1965) and *Rana* (Tata, 1965). As far as evidence goes at present, there is no change in the genome of degenerating tissues: Ryffel, Hagenbüchle and Weber (1973) have shown that the ribosomal DNA, at least, remains unchanged in tail muscle, as in the liver, during all the metamorphic period. Ryffel and Weber (1973) studied in detail the alterations in RNA in several tissues during metamorphosis, however, and found that new ribosomal RNA is formed in most of these tissues. So the activity of the rDNA genes is evidently awakened, although their structure does not change. In the degenerating muscle of the tail, Ryffel and Weber observed a loss of high-molecular-weight RNA.

There has been much less experimental work on organs of *Xenopus* that begin to develop at metamorphosis, than on the regressing tail. Some minor changes in the intestinal epithelium have been described (Bonneville & Weinstock, 1970), and the onset of secretion of 5-hydroxytryptamine (5-HT) in the granular skin glands has been observed by Vanable (1964). There are more general metabolic changes however in which *Xenopus* shows some peculiarities compared with other Anura.

(ii) Nitrogenous Excretion.   When we come to consider the changes in nitrogen metabolism that usually occur at metamorphosis in amphibians, we find that *Xenopus*, unlike other anurans, remains 'aquatic' in these features, for it normally continues to excrete high proportions of ammonia (Munro, 1939). Balinsky and coworkers (1967a,b) have shown, however, that as *Xenopus* adults get older they excrete increasing proportions of urea. They must therefore already have enzymes in the liver that are necessary for urea synthesis. Balinsky's group have shown also that during natural aestivation, when the animals burrow in the mud, as the ponds dry up, or in other conditions tending to cause dehydration (e.g. the addition of saline to the

water), high concentrations of urea accumulate in the tissues. These are excreted in bulk on return to normal conditions. During the dehydration period, carbamyl phosphate synthetase activity increases greatly in the liver tissue. This is a key enzyme in the pathway leading to urea-formation (cf. Figure 10.8), and was regarded by Cohen (1970) as a good indicator of the stage of metamorphosis reached in *Rana*.

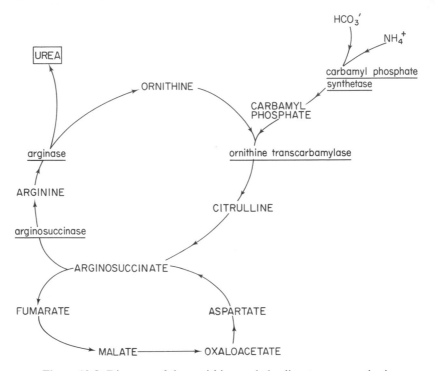

Figure 10.8. Diagram of the ornithine cycle leading to urea synthesis

Janssens (1972) found that *Xenopus* also responded to increased ammonia levels in its environment, by excreting urea in higher proportions. Animals immersed in 5 mM ammonium chloride solution were found to excrete ten times as much urea as normally. Controls in 5 mM sodium chloride showed no such reaction, but according to Henderson and coworkers (1972) the urine they produced was more concentrated and smaller volumes than normal. By contrast, animals kept in distilled water excreted large volumes of very dilute urine. *Xenopus* shows less adaptability than *Rana* in the regulation of urine volume, however.

(iii) Changes in the Blood at Metamorphosis.   There appear, surprisingly, to be no extra oxygen requirements during metamorphosis in *Xenopus*: on the contrary, the oxygen consumption falls (Fletcher & Myant, 1960). There

is of course no sudden change of respiratory mechanism, since the larva was not greatly dependent on its gills for gaseous exchange (cf. Chapters 2 and 5) and already possessed lungs. Probably the efficiency of oxygen binding in the blood increases owing to changes in the type of haemoglobin that is produced at this time. The haemoglobins of *Xenopus* have been studied in some detail by Jurd and McLean (1970). They showed by fluorescent antibody labelling methods that there is a gradual transition from the production of two tadpole haemoglobins, F1 and F2, to the appearance of two new forms, A1 and A2, during late larval stages. The change is complete after metamorphosis, though traces of the tadpole types may persist throughout adult life. McLean and Jurd (1971) and Jurd and McLean (1974) have been able to distinguish which blood cells were producing each haemoglobin, and found that during the transition period, up to 25 % of the cells contained both tadpole and adult types. So evidently the genes responsible for their production can work simultaneously in one cell, for at least a short time. Some cells revert to producing tadpole haemoglobin in cases of anaemia: the reason for this is not clear. Even less is it understood how some rare 'freak' animals reported to have no haemoglobin at all (Ewer, 1959) were able to survive.

There are also changes in the serum proteins at metamorphosis. The albumin/globulin ratio increases and so does the total serum protein (Herner & Frieden, 1960). These trends continue for some time after metamorphosis in *Xenopus*, not stopping abruptly as in Ranidae. Later the serum proteins show more globulins, some of which are antibodies that appear as the immune system develops. We must now say a little more about recent work on both serum and cell-bound antibodies and on the development of immune reactions in *Xenopus*.

4. The Development of the Immune System

There have been a number of recent studies on the development of the immune system in *Xenopus* and its ability to produce both serum antibodies and cell-bound antibodies in response to injections of antigen or grafts of foreign tissue. Cell-bound antibodies are those carried by cells of the lymphocyte system, which cause any cells of another individual to be killed, unless they are genetically identical to the host: so a graft from another individual will be rejected.

Manning and Horton (1969) and Turner (1969) have made a number of observations on the development in *Xenopus* of lymphoid organs which produce the cells that mediate immune reactions. Manning and Horton showed that besides the thymus and spleen, three pairs of lymphoid organs develop in the ventral epithelium of the pharynx, which are homologues of the 'ventral cavity bodies' described by Cooper (1967) in *Rana*. These appear in *Xenopus* larvae at stage 49 (initial hindlimb bud stage), and were described as 'procoracoid bodies' by Sterba (1950). Lymphoid tissue also appears in the liver, the mesonephric kidney, and a very little in the gut wall, but there is none in the pronephros and there are no lymph nodes in *Xenopus*, unlike other anurans. The differentiation of the thymus slightly precedes that of other lymphoid organs and it contains small lymphocytes—the cells responsible for producing antibodies—by stage 49. DuPasquier, Weiss and Loor (1972) have in more recent studies confirmed that the thymus is the first source of lymphocytes with surface-borne immunoglobulins, in *Xenopus* larvae.

The time of onset of cellular immune reactions in *Xenopus* was followed by Horton (1969) and Manning (1971) by exchanges of skin grafts between larvae and frogs of various ages. They found that whatever the larval stage at which the graft was made, a graft-rejection reaction set in at the time when the thymus first contained mature lymphocytes. Since then Kidder, Ruben and Stevens (1973) have used a more sensitive method of recognizing immune reactions, in tissue culture, and have found that spleen cells too can mediate an immune response to foreign tissue, by stage 50 in *Xenopus* larvae. This is when small lymphocytes first appear in the spleen of the intact animal. Turner and Manning (1973) have now studied cellular proliferation in the spleen in response to antigen injections. They find no evidence of the existence of 'plasma cells' similar to those found in mammals during the immune response. Another point of interest is that even after removal of the spleen, *Xenopus* is capable of producing serum antibodies (Turner, 1973). So these antibodies evidently come from sources other than the spleen.

In their work with skin grafts, Manning and Horton found no evidence of induction of 'tolerance' in the host: i.e. when the grafts were made some time before the immune system of the host was mature, this did not enable them to survive as if they had been accepted as the host's own tissue. This raises the question of how it is that grafts made into embryos are able to survive. Most of experimental embryology has only been possible because of this convenient fact. The reason may be partly that the results of such experiments are often harvested at young larval stages, before the allograft reaction could set in. Another point, observed by Volpe and Gebhardt (1965) and by Horton (1969) is that the size of the graft determines its survival time. If the graft is large relative to the host embryo, as it often is in experiments on early stages, it may perhaps sequester the early lymphocytes of the host as fast as these are produced, and so knock out any possible allograft reaction. If the graft itself contains lymphocytes, it will occasionally mount an immune reaction against the host: this sometimes happens with grafts of spleen tissue (DeLanney, 1958).

It is in fact possible to induce tolerance to foreign tissues in *Xenopus*, as Clark and Newth (1972) have shown. Their method was to place epidermal grafts onto late neurula stages. They have also shown that the spleen is involved in cellular immune reactions, since grafts of spleen from immunized animals into tolerant hosts caused the hosts to reject skin grafts that they had previously tolerated. At the same time the spleen was able to mount a graft-versus-host reaction. Ruben, Stevens and Kidder (1972) have produced evidence that spleen grafts are also capable of suppressing the normal immune reactions of a host. If spleen, liver and kidney from the same donor animal are grafted into a *Xenopus* larva, it gives weaker, delayed rejection reactions to the liver and kidney, compared with controls which have no spleen grafted as well. The mechanism of this suppression of the host's immune reaction by the grafted spleen is not understood fully yet.

The production of serum antibodies in *Xenopus* has so far been investigated mainly in adults (see Jurd & Stevenson, 1974), although Jurd (1972) is now turning attention to larval stages and has shown that at stage 47 (when all yolk has been absorbed) animals are already capable of producing these antibodies in response to antigenic stimulation. Earlier investigators (for example Butler, Rees and Elek, 1962; Elek, Rees and Gowing, 1962) who worked on *Xenopus*, had doubted the ability of cold-blooded tetrapod vertebrates to mount the same sequence of reactions as in mammals. It was soon found, however, that a '19S' class of antibodies, of relatively high molecular weight, comparable to the γM immunoglobulins of mammals, could be detected in common frogs (*Rana* spp.) and toads (*Bufo* spp.) within a few days after injection of antigen. It was then found that after a few weeks, a low-molecular-weight, '7S' class of antibody could be detected, comparable with the γG immunoglobulins of mammals, though this did not always entirely replace the 19S type as it does in the rabbit.

There have been recent studies of the serum antibodies produced in adult *Xenopus* in response to a variety of antigens (see Lykakis & Cox, 1968; Lykakis, 1969; Marchalonis, Allen & Saarni, 1970; Hadji-Azimi, 1971; Yamaguchi *et al.* 1973). The antigens used have included bovine γ-globulin, haemocyanins from invertebrates, and the bacterium *Salmonella flagella*. In response to any of these, adult *Xenopus* will produce both 19S and 7S serum antibodies. The 19S antibody appears first, and after about 4 weeks in some cases, 7S antibody takes its place. In larvae too, Jurd has been able to show that 19S antibody is produced.

When one thinks again about the timing of the onset of these immune reactions in *Xenopus*, it is surprising that they take place so long before the process of metamorphosis is complete. Many new tissues form in the limbs, and many new enzymes in the gut and liver, after the immune system has become able to reject 'foreign' or unfamiliar cells and proteins. One wonders how these new tissues and enzymes avoid being rejected by their own immune cells. It may be that, like the large grafts on embryos, they present too much for the available population of lymphocytes to cope with. Or perhaps they continue to carry sufficient 'marker' antigens from the larval organs that they supersede, for the immune system to recognize them as 'self' and not foreign. We have no evidence on either of these possibilities so far: it seems a wide-open and fascinating field for further research.

Perhaps one of the most remarkable features of this fascinating process of metamorphosis, however, is that it so rarely fails under natural conditions. In fact, if it is suppressed experimentally for long periods, this can result in abnormal expressions of the potentialities for rapid growth that are apparently innate in the larval tissues. For instance, Pflugfelder (1959) obtained tumour-like growths in the skin, gut, gonads and mesenchyme of *Xenopus* larvae, when metamorphosis was blocked by treating them with potassium chlorate. The ability of these larval tissues to grow, regress, regenerate, or form tumours in different conditions emphasizes their remarkable versatility, which must result from properties of the genome about which we know virtually nothing yet. Larval amphibian cells could well be exploited for cancer research and for genetical studies, much more than they have been in the past.

11

Future Possibilities in Research on *Xenopus*

It is not only rash, but also unscientific for anyone to claim to be able to predict the future trends of research and discovery in any branch of biology. In research, no one can know exactly what the findings will be, or what new ideas they may engender. Each investigator can select quite narrowly what kinds of result he will look for, however, thus keeping the main trend of his own investigations in one direction (and thereby missing many other interesting possibilities). So one can often predict what kinds of phenomena certain investigators will look for next, if they continue in the lines along which they have already started. An outside reviewer may also be able to see promising new lines that could profitably be followed in the future, while the active investigators are too absorbed in their own special interests to have seen these new possibilities yet.

In the last half of this chapter I have picked out a few of the most recent reports of work on *Xenopus* that seem to promise healthy growing-points for future research, provided that certain needs are met in techniques and in the supplies of animals. Before considering these, let us first summarize briefly some of the main advances in biological knowledge that have resulted from work on *Xenopus* up to date.

The early work on the anatomy of *Xenopus* was both helpful and a source of controversy for students of the comparative anatomy of amphibians. It helped to establish the relationship of the Pipidae to other aglossan amphibians and to the rest of the Anura (cf. Table 1.2, p. 19). But *Xenopus*'s several similarities to urodeles forced the experts to reconsider which characters in amphibians should be regarded as primitive and which as specializations for aquatic life. We do not yet have certain answers to these points. As we saw in Chapter 5, there are also peculiar features about the anatomy of the *Xenopus* larva: its lack of efferent branchial arteries, its large lungs, and its filter-feeding system. These features serve to emphasize that a wide range of larval adaptations can occur quite independently of adult characters. There are doubtless many other examples of larval specializations in the Amphibia, but the problem of larval evolution in this class of animals has not received nearly so much attention as it has in some of the invertebrate phyla. Theories of vertebrate evolution might gain a great deal from a more detailed consideration of the larval as well as the adult forms of amphibians, and of genetic variants of these.

Physiological and endocrinological work on *Xenopus* has led to many important findings: for instance on the action of the melanophore-stimulating hormone and the mechanism of colour change in response to background, and on the roles of gonadotrophic hormones in lower vertebrates. The activities of pituitary cells in coordinating the events both of reproduction and of metamorphosis in amphibians have become clearer through work on *Xenopus*. In addition, many examples have been seen of the remarkable specificity in the action of thyroxine on different tissues during metamorphosis. But the reasons for these different effects of thyroxine remain unknown so far. Now that more techniques are available for the detection of enzyme activities and other biochemical changes in cells in tissue culture, it should be possible to investigate in more detail what are the immediate metabolic responses of each tissue *in vitro* to thyroxine. As we saw in Chapter 10, Weber and his colleagues have been able to show several enzymatic changes in tail tissue of *Xenopus* larvae in response to thyroxine.

It was also pointed out in the last chapter that a number of investigations are still in train on the development of immune mechanisms in *Xenopus*, before and after metamorphosis. Jurd's survey of the appearance of serum antibodies in larval stages should soon be complete, and work is being carried on by both Manning and Horton and their collaborators, and DuPasquier and his team, on the cellular immune responses. In the course of this work it may become clear why the newly-developed adult tissues which appear at metamorphosis do not arouse any auto-immune response in their own cellular immune system.

Besides its immunology, the neuroendocrinology of *Xenopus* offers promising possibilities for future research, on the basis of important discoveries already made. Little is known yet about the extent to which regeneration and metamorphosis, as well as reproduction, are controlled in amphibia by the hypothalamus of the brain. Hauser's observations on the role of Reissner's fibres in controlling regeneration of the tail (see Chapter 10) indicate that in *Xenopus* as in other lower vertebrates, the subcommissural organ may be another important source of neurosecretion. So far, TRF-secreting cells have been identified in the hypothalamus of *Xenopus* (Goos, 1969), suggesting that the release of thyroid hormone is under hypothalamic control just as it is in mammals. The small size of the amphibian brain compared with that of mammals has made it impossible so far to study the functions of hypothalamic neurosecretory cells in the same detail as has been done in mammals by means of microelectrodes. The possibility of more accurate mapping of the functions of different hypothalamic areas in amphibians is something to be borne in mind for the future, however. Gaze and Jacobson and their associates achieved very precise mapping of such small regions as the optic tectum and the retina, in *Xenopus*. Their work on retinotectal connections represents another outstanding advance. Further studies on the development of retinotectal connections in *Xenopus* have been made by Gaze *et al.* (1972), Chung *et al.* (1973), Lázár (1973) and Scott (1974). What are now much needed are experiments to test Gaze's new concept of 'systems matching' between retinal and tectal areas.

Another important group of findings on the relationship between the nervous system and end organs in development are those of Hughes and his colleagues (cf. Chapter 10). They were the first to show that sensory as well as motor neurons were affected by the presence or absence of limbs, and to estimate the numbers of neuroblasts that degenerate, both in normal development and after limb ablation.

It is in the field of developmental biochemistry that work on *Xenopus* has produced some of the most notable advances in knowledge recently. The impetus to this work was given by the discovery of the anucleolate mutant by Elsdale, Fischberg and Smith (1958): this enabled the essential role of nucleoli in development to be seen clearly. It also led to the observation that even in normal embryos, no nucleoli and no new ribosomes form until just before the gastrula stage: i.e. that maternal ribosomes suffice for all the syntheses in early development. It was also confirmed that in *Xenopus*, as in *Rana* and in sea-urchins, little or no DNA or RNA are synthesized during cleavage stages, although the number of nuclei is increasing very rapidly. We still do not know what is the source of the new nuclear DNA during cleavage. It seems most likely that it is derived from yolk-bound cytoplasmic DNA (cf. Chapter 6), but this has not yet been proved by any labelling experiments. It should be possible to establish this now, however, since autoradiographic work combined with electron microscopy is a routine in most laboratories.

Nuclear transplantation work has been a major step forward in research technique, and is another line of investigation in which work on *Xenopus* has led the field recently. This was again due to the impetus given by the discovery of the anucleolate mutant, which made it possible to distinguish transplanted nuclei from that of the host egg with certainty, if the host is a mutant and the transplanted nuclei are normal. In *Xenopus*, Gurdon and his associates have demonstrated the special role of egg cytoplasm in promoting replication of nuclei, and even of purified DNA. Quite unusual nuclei may be stimulated to divide in *Xenopus* egg cytoplasm, as we shall see when some of the most recent work is considered below (see also Brun, 1973).

Reports of new findings on *Xenopus* appear continuously in the current scientific journals, and it would be impossible to give a completely up-to-date review of them. Just a few representatives of continuing lines of work and of recent innovations have been selected in the following paragraphs.

Detailed work on early and late morphogenesis of tissues and organs is still being pursued in *Xenopus*, with the aid of the electron microscope and of cytochemical methods. These tools allow some insight into physiological events in the cells too. An example of this approach is Meban's description (1973) of the cells lining the surface of the lung in *Xenopus*. These 'pneumonocytes' have long cytoplasmic processes which extend over the surface of the capillary blood vessels that supply the lung. They also contain osmiophil bodies in the cytoplasm and many microvilli at the free surface of the cell. On the basis of these findings, Meban suggests that the pneumonocytes function not only in gas exchanges but also in secretion, perhaps of surface-active substances such as are found in the alveoli of the mammalian lung.

Tissue interaction in the early embryo—particularly the problem of neural induction—still draws many investigators. Tarin (1973) has pursued his investigation of the material lying between the inducing mesoderm and the dorsal ectoderm in *Xenopus* gastrulae, and he concludes from histochemical tests that the granular material consists mainly of RNA, while certain filaments which appear later contain mucopolysaccharides. He finds, however, that digestion of the material with RN-ase and amylase does not prevent neural induction from occurring. So he has concluded that this material plays no essential role in the interaction between mesoderm and ectoderm. It seems possible, however, that the material could already have fulfilled its role by the stage at which Tarin's digestion procedures took place. His method of introducing the enzymes, by microinjection into the blastocoel, may not have allowed a sufficiently rapid destruction of the intercellular materials. It might be better to use cultures of smaller number of cells and to treat mesoderm alone with the enzymes before combining it with ectoderm. As we saw in Chapter 8, however, a minimal number of at least 100 ectoderm cells should be used, to ensure that proper neural differentiation is possible.

Recently Cooke (1973) has refocused attention on the numbers of cells required for normal differentiation in amphibian embryos. He has found, surprisingly, that both gastrulation and neurulation can proceed normally, and that all the larval tissues will differentiate, even if mitosis has been completely blocked from the early gastrula stage. It is most remarkable that the normal patterns and groupings of cells into tissues, as well as normal morphogenetic movements, can occur even when much smaller numbers of cells are present than normal. Cooke sees this as an example of the ability of embryonic cells to exchange 'positional information' (Wolpert *et al.*, 1971) so that they can adapt their behaviour and fates according to their numbers and relative positions. Although embryologists know many examples of this ability of embryos to adapt and develop normally in the face of adverse circumstances, it still remains a puzzle how they evolved this adaptability, since their natural environment in the past certainly did not include hazards from experimenters!

In Chapter 7 we saw that attempts have been made to explain the mechanisms of interaction of embryonic cells in terms of their surface properties and mobility. Further research with the electron microscope should enable us to see more of their surface structure: so far most of such studies have been on cleavage stages. Recently Bluemink and his colleagues have been pursuing their investigations on the effect of Cytochalasin B on microfilaments and microtubules in cleavage cells of *Xenopus* (DeLaat, Luchtel & Bluemink, 1973). They have shown that the agent is more effective when injected into the cells, and also that the membrane potential of the cells rises at the time when they are most sensitive to cytochalasin.

It is difficult to judge whether future work on the electrophysiological properties of embryonic cells will lead to any major breakthrough in our understanding of the ways in which cells interact. It has been a remarkable achievement already, to show that cells other than neurons can pass on electrical potential changes. As techniques for detecting these changes improve, it may become possible to recognize differing patterns of electrical discharges, which could be interpreted as 'signals' for different chains of events in the cells.

Biochemical work on *Xenopus* continues: notably, Kunz (1973) has extended the study of lactate dehydrogenase isozymes which she started with Hearn in the 1960's. She has now shown the complete pattern of development of isozyme components, in different regions of the tailbud embryo and in several organs from larval to adult stages. More recently, Wall and Blackler (1974) have shown that it is possible to distinguish differences in the subunits of malate dehydrogenase (MDH) isozymes, in embryos of different species of *Xenopus* and in their hybrids. This work opens possibilities of observing single gene differences between the species.

Research on nuclear activity in *Xenopus* embryos will be a main line of interest for some time, since so much valuable groundwork has been provided by the earlier nuclear transplantation experiments and the studies on nucleoli and ribosomal genes. Recently Kobel, Brun and Fischberg (1973) reported successful nuclear transplants from epidermal cells of larvae, from melanophores, and from a cultured line of aneuploid adult liver cells called A-8. They found that nuclei from ciliated epidermal cells did not allow development beyond blastula stage, but that non-ciliated epidermis nuclei could support normal development up to larval stages. Melanophore nuclei from cultured lines of cells gave more successful development than did those taken direct from larvae. Surprisingly, the aneuploid A-8 cell nuclei were also capable of supporting development of a normal larva. It is possible that the cultured nuclei are successful because they have been preselected for viability during the culture period: but it is still surprising that aneuploids should do so well. We know very little about what parts of the gene and chromosome complement are essential for normal embryonic development, since so far the only defective nuclei identified in *Xenopus* are aneuploids and nucleolar mutants. We know hardly anything yet about the activities of single genes in this animal.

It is in the genetics of *Xenopus* that there seems now the greatest need for further work. A brilliant start has been made in its biochemical genetics, by Brown and his associates who have recently analysed the structures of the ribosomal genes in different *Xenopus* species (see Brown, 1973; Brown, Wensinck & Jordan, 1972; Brown & Sugimoto, 1973). They have shown that the DNA units coding for 5S, 18S and 28S ribosomal RNA are identical in the two species, *Xenopus laevis* and *X. mulleri*, but that there are different base sequences in the 'spacer' regions between the genes, in the two species. It would be interesting to know to what extent these base sequences are identical in different individuals of the same species. This cannot be found out until it is possible to compare samples of embryos from different pure, homozygous inbred lines of animals. Nace (personal communication 1973) is now attempting to build up some inbred stocks of *Xenopus laevis* at the University of Michigan, so that its genetics can be studied more precisely in the future. It is to be hoped that some easily-recognizable colour patterns will be found to depend on simple gene differences, so that these can be used as 'markers' for different inbred strains, like the coat-colours of mice. However, the slow rate at which *Xenopus* individuals become sexually mature (18 months to two years, in the laboratory) makes them far more difficult material than mice for genetical work and for the establishment of pure-line laboratory stocks.

It is not known to what extent populations of *Xenopus* become inbred in the wild. Small, isolated populations of some of the rarer species may do so, but *Xenopus laevis* is far too abundant and widespread for extensive inbreeding to be likely. From Nace's personal report, there is little danger that the stocks of *X. laevis* in South Africa will become depleted by demands for them for biological research.

There is, however, another possible risk to the survival of wild populations of *Xenopus* in Africa: I am told that they *taste* delicious, to those who have no prejudices against eating fried frogs! Is it possible, then, that *Xenopus* may become less readily available for research, as the world shortage of protein becomes more acute, and that they may find their way more often to the dinner-table than to the laboratory? Ugh! Let us hope that man's intellectual needs will always take precedence over his appetite, where *Xenopus* is concerned! There is much more that we can and should learn from laboratory studies of this unique, aquatic anuran.

Bibliography

Note: some recent references, added in proof, are listed separately on p. 233

Acher, R., Beaupain, R., Chauvet, J., Chauvet, M.T. & Crepy, D. (1964). The neurohypophyseal hormones of the amphibians: comparison of the hormones of *Rana esculenta* and *Xenopus laevis. Gen. comp. Endocrinol.*, **4**, 596–601.

Adam, H. (1954). Freie kugelförmige Pigmentzellen in den Gehirnventrikeln von Krallenfroschlarven *Xenopus laevis* (Daudin). *Z. mikrosk.- anat. Forsch.*, **60**, 6–32.

Ahl, E. (1924). Ueber einige afrikanische Frosche. *Zool. Anz.*, **60**, 270–1.

Alexander, S.S. & Bellerby, C.W. (1935). The effects of captivity upon the reproductive cycle of the South African Clawed Toad, *Xenopus laevis. J. exp. Biol.*, **12**, 306–14.

Al-Mukhtar, K.A.K. & Webb, A.C. (1971). An ultrastructural study of primordial germ cells, oogonia and early oocytes in *Xenopus laevis. J. Embryol. exp. Morph.*, **26**, 195–217.

Amaldi, F., Lava-Sanchez, P.A. & Buon-Giorno-Nardelli, M. (1973). Nuclear DNA content variability in *Xenopus laevis*: a redundancy regulation common to all gene families. *Nature, Lond.*, **242**, 615–17.

Ambellan, E. & Webster, G. (1962). Effects of nucleotides on neurulation in amphibian embryos. *Devl. Biol.*, **5**, 452–67.

Andrew, A. & Gabie, V. (1969a). Hanging-drop culture of *Xenopus laevis* neural crest. *Acta Embryol. exp.*, pp. 123–6.

Andrew, A. & Gabie, V. (1969b). Staging of pigment cells in cultures of *Xenopus laevis* neural crest. *Acta Embryol. exp.*, pp. 137–46.

Ansell, G.B. & Richter, D. (1954). A note on the free amino acid contents of brain. *Biochem. J.*, **57**, 70–3.

Arnoult, J. & Lamotte, M. (1968). Les Pipidae de l'Ouest africain et du Cameroun. *Bull. inst. fr. Afr. noire.*, **30**(A), 270–306.

Atwell, W.J. (1941). The morphology of the hypophysis cerebri of toads. *Amer. J. Anat.*, **68**, 191–208.

Ave, K., Kanakami, I. & Sameshima, M. (1968). Studies on the heterogeneity of cell populations in amphibian presumptive epidermis with reference to primary induction. *Devl. Biol.*, **17**, 617–26.

Babcock, R.G. (1963). RNA and lens induction in *Xenopus laevis. Amer. Zoologist.*, **3**, 511–12.

Bachvarova, R., Davidson, E.H., Allfrey, V.G. & Mirsky, A.E. (1966). Activation of RNA synthesis associated with gastrulation. *Proc. natn. Acad. Sci. U.S.A.*, **55**, 358–65.

Backström, S. (1954). Morphogenetic effects of lithium on the embryonic development of *Xenopus*. *Ark. Zool.*, **6**, 527–36.

Backström, S. (1962). A link between neural induction and vegetalization. The effect of magnesium on *Xenopus* embryos. *Acta Embryol. Morph. exp.*, **5**, 295–303.

Bagnara, J.T. (1960). Tail melanophores of *Xenopus* in normal development and regeneration. *Biol. Bull. mar. biol. lab. Woods Hole.*, **118**, 1–8.

Bagnara, J.T. (1963). The pineal and the body lightening reaction of larval amphibians. *Gen. comp. Endocrinol.*, **3**, 86–100.

Baker, P.C. (1965). Fine structure and morphogenetic movements in the gastrula of the tree frog, *Hyla regilla*. *J. Cell Biol.*, **24**, 95–116.

Baker, P.C. (1966). Monoamine oxidase in the eye, brain and whole embryo of developing *Xenopus laevis*. *Devl. Biol.*, **14**, 267–77.

Baker, P.C. & Hoff, K.M. (1971). Melatonin localization in the eyes of larval *Xenopus*. *Comp. Biochem. Physiol.*, **39A**, 879–91.

Baker, P.C. & Schroeder, T.E. (1967). Cytoplasmic filaments and morphogenetic movements in the amphibian neural tube. *Devl. Biol.*, **15**, 432–50.

Balinsky, B.I. (1958). On the factors controlling the size of the brain and eyes in anuran embryos. *J. exp. Zool.*, **139**, 403–38.

Balinsky, B.I. (1960). Ultrastructural mechanisms of gastrulation and neurulation. In: *Symposium on Germ Cells and Early Stages of Development* (ed. S. Ranzi), pp. 550–63. Milan: Fond. Baselli.

Balinsky, B.I. (1966). Changes in the ultrastructure of amphibian eggs following fertilization. *Acta Embryol. Morph. exp.*, **9**, 132–54.

Balinsky, B.I. (1970). *An Introduction to Embryology* (3rd edn.). Philadelphia, London, Toronto: W.B. Saunders & Co.

Balinsky, B.I. & Devis, R.J. (1963). Origin and differentiation of cytoplasmic structures in the oocytes of *Xenopus laevis*. *Acta Embryol. Morph. exp.*, **6**, 55–108.

Balinsky, J., Choritz, E.L., Coe, C.G.L. & Van der Schans, G.S. (1967a). Urea cycle enzymes and urea excretion during the development and metamorphosis of *Xenopus laevis*. *Comp. Biochem. Physiol.*, **22**, 53–7.

Balinsky, J., Choritz, E.L., Coe, C.G.L. & Van der Schans, G.S. (1967b). Amino acid metabolism and urea synthesis in naturally aestivating *Xenopus laevis*. *Comp. Biochem. Physiol.*, **22**, 59–68.

Balls, M. & Ruben, L.N. (1964). Variations in the response of *Xenopus laevis* to normal tissue homografts. *Devl. Biol.*, **10**, 92–104.

Balls, M. & Ruben, L.N. (1967). The transmission of lymphosarcoma in *Xenopus laevis*, the South African Clawed Toad. *Cancer Res.*, **27**, 654–9.

Baltus, E. & Brachet, J. (1962). Le dosage de l'acide désoxyribonucléique dans les oeufs de Batraciens. *Biochim. biophys. Acta.*, **61**, 157–63.

Barr, H.J. (1966). Problems in the developmental cytogenetics of nucleoli in *Xenopus*. *Nat. Cancer Inst. Monogr.*, **23**, 411–24.

Barth, L.G. & Barth, L.J. (1962). Further investigation of the differentiation *in vitro* of presumptive epidermis cells of the *Rana pipiens* gastrula. *J. Morph.*, **110**, 347–73.

Beddard, F.E. (1894). Notes upon the tadpole of *Xenopus laevis* (*Dactylethra capensis*). *Proc. Zool. Soc. Lond.*, pp. 101–7.

Bellairs, R. (1963). The development of somites in the chick embryo. *J. Embryol. exp. Morph.*, **11**, 697–714.

Bellerby, C.W. & Hogben, L. (1938). Experimental studies on the sexual cycle of the South African Clawed Toad, *Xenopus laevis*. *J. exp. Biol.*, **15**, 91–100.

Bentley, P.J. (1969). Neurohypophysial hormones in Amphibia: a comparison of their actions and storage. *Gen. comp. Endocrinol.*, **13**, 39–44.

Benz, G. (1957). Regionale Verteilung der Kathepsinaktivität im Schwanz von gefütterten und hungernden *Xenopus*-Larven. *Rev. Suisse Zool.*, **64**, 337–49.

Berk, L. (1938). Studies on the reproduction of *Xenopus laevis*. I. The relation of external environmental factors to the sexual cycle. *S. Afr. J. med. Sci.*, **3**, 72–7.

Berk, L. (1939). Studies on the reproduction of *Xenopus laevis*. III. The secondary sex characters of the male *Xenopus*: the pads. *S. Afr. J. med. Sci.*, **4**, 53–60.

Berk, L., Cheetham, R.W.S. & Shapiro, H.A. (1936). The biological basis of sexual behaviour in Amphibia. III. The role of distance receptors in the establishment of the mating reflex (coupling) in *Xenopus laevis* (the South African Clawed Toad): the eyes. *J. exp. Biol.*, **13**, 60–2.

Berk, L. & Shapiro, H.A. (1939). Studies on the reproduction of *Xenopus laevis*. II. The histological changes in the accessory sex organs of female *Xenopus* induced by the administration of endocrine preparations. *S. Afr. J. Med. Sci.*, **4**, 13–17.

Bijtel, J.H. (1931). Ueber die Entwicklung des Schwanzes bei Amphibien. *Wilhelm Roux Arch. EntwMech. Org.*, **125**, 448–86.

Birnstiel, M., Grunstein, M., Speirs, J. & Hennig, W. (1969). Family of ribosomal genes in *Xenopus laevis*. *Nature, Lond.*, **223**, 1265–7.

Blackler, A.W. (1958). Contribution to the study of germ cells in the Anura. *J. Embryol. exp. Morph.*, **6**, 491–503.

Blackler, A.W. (1960) Transfer of germ cells in *Xenopus laevis*. *Nature, Lond.*, **185**, 859.

Blackler, A.W. (1962). Transfer of primordial germ cells between two subspecies of *Xenopus laevis*. *J. Embryol. exp. Morph.*, **10**, 641–51.

Blackler, A.W. & Fischberg, M. (1968). Hybridization of *Xenopus laevis petersi (poweri)* and *X. l. laevis*. *Rev. Suisse Zool.*, **75**, 1023–32.

Blackler, A.W., Fischberg, M. & Newth, D.R. (1965). Hybridization of two subspecies of *Xenopus laevis* (Daudin). *Rev. Suisse Zool.*, **72**, 841–57.

Blackler, A.W. & Gecking, C.A. (1972). Transmission of sex cells of one species through the body of a second species within the genus *Xenopus*. *Devl. Biol.*, **27**, 376–85.

Bles, E.J. (1905). The life history of *Xenopus laevis* Daud. *Trans. R. Soc. Edinb.*, **41**, 789–850.

Bluemink, J.G. (1971a). Effects of cytochalasin B on surface contractility and cell junction formation during egg cleavage in *Xenopus laevis*. *Cytobiologie.*, **3**, 176–87.

Bluemink, J.G. (1971b). Cytokinesis and cytochalasin-induced furrow regression in the first-cleavage zygote of *Xenopus laevis*. *Z. Zellforsch.*, **121**, 102–26.

Bocage, B.du (1895). *Herpétologie d'Angola et du Congo.* p. 187.

Bogenschütz, H. (1965). Extraokulare Steuerung des Farbwechsels bei Kaulquappen. *Experientia.*, **21**, 451.

Boie, H. (1827). Letter in *Isis*, **20**, 726.

Bonneville, M.A. & Weinstock, M. (1970). Brush border development in the intestinal absorptive cells of *Xenopus* during metamorphosis. *J. Cell Biol.*, **24**, 151–71.

Borradaile, L.A. (1941). *A Manual of Elementary Zoology*, Oxford University Press.

Bosman, L.P. & Zwarenstein, H. (1930). The effect of temperature on the carbohydrate tolerance in *Xenopus laevis* (South African Clawed Toad). *Q. Jl. exp. Physiol.*, **20**, 231–43.

Bosman, L.P. & Zwarenstein, H. (1934). The blood sugar level in normal and eyeless *Xenopus laevis*. *Trans. Roy. Soc. S. Afr.*, **22**, xviii.

Bounoure, L. (1934). Recherches sur la lignée germinale chez la grenouille rousse aux premiers stades du développement. *Annls. Sci. nat.*, **10e**, ser. **17**, 67–248.

Bradley, S. (1970). An analysis of self-differentiation of chick limb buds in chorio-allantoic grafts. *J. Anat. Lond.*, **107**, 479–90.

Brachet, J. (1961). Morphogenetic effects of lipoic acid on amphibian embryos. *Nature, Lond.*, **189**, 156–7.

Brachet, J. (1968). Emission of Feulgen-positive particles during the *in vitro* maturation of toad ovocytes. *Nature, Lond.*, **208**, 596–7.

Brachet, J., Hanocq, F. & Van Gansen, P. (1970). A cytochemical and ultrastructural analysis of *in vitro* maturation in amphibian oocytes. *Devl. Biol.*, **21**, 157–95.

Brachet, J. & Malpoix, P. (1971). Macromolecular syntheses and nucleocytoplasmic interactions in early development. *Adv. Morphogen.*, **9**, 263–316.

Brahma, S.K. (1958). Experiments on the diffusibility of the amphibian evocator. *J. Embryol. exp. Morph.*, **6**, 418–23.

Brahma, S.K. (1959). Studies on the process of lens induction in *Xenopus laevis*. *Wilhelm Roux Arch. EntwMech. Org.*, **151**, 181–7.

Briggs, R., Green, E.U. & King, T.J. (1951). An investigation of the capacity for cleavage and differentiation in *Rana pipiens* eggs lacking functional chromosomes. *J. exp. Zool.*, **116**, 455–61.

Briggs, R. & King, T.J. (1952). Transplantation of living nuclei from blastula cells into enucleated frogs' eggs. *Proc. natn. Acad. Sci. U.S.A.*, **38**, 455–63.

Bristow, D.A. & Deuchar, E.M. (1964). Changes in nucleic acid concentration during the development of *Xenopus laevis* embryos. *Expl. Cell Res.*, **35**, 580–9.

Brocas, J. & Verzar, F. (1961). The ageing of *Xenopus laevis*. *Experientia*, **17**, 421.

Brouwer, E. (1970). The involvement of catecholamines in the dispersion reaction of the melanophores of *Xenopus laevis* in vivo. *Gen. comp. Endocrinol.*, **15**, 264–71.

Brouwer, E. (1972). A chemical and pharmacological study on the role of catechol-amines in the dispersion reaction of *Xenopus laevis*. *Gen. comp. Endocrinol.*, **18**, 378–83.

Brown, D.D. (1973). The isolation of genes. *Scientific American*, **229**, 20–9.

Brown, D.D. & Gurdon, J.B. (1966). Size distribution and stability of DNA-like RNA synthesized during development of anucleolate embryos of *Xenopus laevis*. *J. molec. Biol.*, **19**, 399–422.

Brown, D.D. & Littna, E. (1964). RNA synthesis during the development of *Xenopus laevis*, the South African Clawed Toad. *J. molec. Biol.*, **8**, 669–87.

Brown, D.D. & Sugimoto, K. (1973). 5S DNAs of *Xenopus laevis* and *Xenopus mulleri*: evolution of a gene family. *J. molec. Biol.*, **78**, 397–416.

Brown, D.D., Wensink, P.C. & Jordan, E.A. (1972). Comparison of the rDNA of *Xenopus laevis* and *X. muelleri*. *J. molec. Biol.*, **63**, 57–73.

Brun, R. (1973). Mammalian cells in *Xenopus* eggs. *Nature (New Biol.) Lond.* **243**, 26–7.

Buehr, M.L. & Blackler, A.W. (1970). Sterility and partial sterility in the South African Clawed Toad following the pricking of the egg. *J. Embryol. exp. Morph.*, **23**, 375–84.

Burgers, A.C.J. (1952). Optomotor reactions of *Xenopus laevis*. *Physiol. comp. & oecol.*, **2**, 272–81.

Burgers, A.C.J. (1956). Investigations into the action of certain hormones and other substances on the melanophores of the South African Clawed Toad, *Xenopus laevis*. Proefschrift für Ph.D., Arnhem: G.W. Van der Wiel & Co.

Burgers, A.C.J., Imai, K. & Van Oordt, G.J. (1963). The amount of melanophore-stimulating hormone in single pituitary glands of *Xenopus laevis* kept under various conditions. *Gen. comp. Endocrinol.*, **3**, 53–7.

Burgess, A.M.C. (1967). The developmental potentialities of regeneration blastema nuclei as determined by nuclear transplantation. *J. Embryol. exp. Morph.*, **18**, 27–41.

Butler, L.O., Rees, T.A. & Elek, S.D. (1962). Studies on the serum proteins of *Xenopus laevis* Daudin. *Comp. Biochem. Physiol.*, **6**, 105–10.

Callaghan, O.H. (1953). The effect of temperature on the active ion-transport mechanism and respiration of the skin of *Xenopus laevis.*, *S. Afr. J. med. Sci.*, **18**, 34.

Callan, H.G. (1952). A general account of experimental work on amphibian oocyte nuclei. *Brit. Soc. exp. Biol. Symposium*, **VI**, pp. 243–55.

Callan, H.G. & Tomlin, S.G. (1950). Experimental studies on amphibian oocyte nuclei. I. Investigation of the structure of the nuclear membrane by means of the electron microscope. *Proc. Roy. Soc. B*, **137**, 367–78.

Campbell, J.C. (1963). Lens regeneration from iris, retina and cornea in lentectomised eyes of *Xenopus laevis. Anat. Rec.*, **145**, 214.

Campbell, J.C. (1965). An immuno-fluorescent study of lens regeneration in larval *Xenopus laevis. J. Embryol. exp. Morph.*, **13**, 171–9.

Chang, C.Y. (1953). Parabiosis and gonad transplantation in *Xenopus laevis* Daudin. *J. exp. Zool.*, **123**, 1–28.

Chang, C.Y. & Witschi, E. (1965). Breeding of sex-reversed males of *Xenopus laevis* Daudin. *Proc. Soc. exp. Biol. Med.*, **89**, 150–2.

Chang, C.Y. & Witschi, E. (1956). Genic control and hormonal reversal of sex differentiation in *Xenopus. Proc. Soc. exp. Biol. Med.*, **93**, 140–4.

Charlton, H.M. (1966). The pineal gland and colour change in *Xenopus laevis* Daudin. *Gen. comp. Endocrinol.*, **7**, 384–97.

Charlton, H.M. (1968). The pineal gland of *Xenopus laevis* Daudin: a histological, histochemical and electron-microscopical study. *Gen. comp. Endocrinol.*, **11**, 465–80.

Clark, J.C. & Newth, D.R. (1972). Immunological activity of transplanted spleens in *Xenopus laevis. Experientia*, **28**, 951–3.

Claycomb, W.C. & Villee, C.A. (1971). Lactate dehydrogenase isozymes of *Xenopus laevis:* factors affecting their appearance during early development. *Devl. Biol.*, **24**, 413–27.

Clayton, R.M. (1953). Antigens in the developing newt embryo. *J. Embryol. exp. Morph.*, **1**, 25–42.

Coggins, L.W. (1973). An ultrastructural study of early oogenesis in the toad *Xenopus laevis. J. Cell Sci.*, **12**, 71–94.

Coghill, G.E. (1914). Correlated anatomical and physiological studies of the growth of the nervous system in Amphibia. *J. comp. Neurol.*, **24**, 161–234.

Cohen, A.G. (1967). Observations on the pars intermedia of *Xenopus laevis. Nature, Lond.*, **215**, 55–6.

Cohen, M.I. & Morrill, G.A. (1969). Model for the electric field generated by unidirectional sodium transport in the amphibian embryo. *Nature, Lond.*, **222**, 84–6.

Cohen, P.P. (1970). Biochemical differentiation during amphibian metamorphosis. *Science, N.Y.*, **168**, 533–43.

Coleman, R., Evennett, P.J. & Dodd, J.M. (1968). Ultrastructural observations on the thyroid gland of *Xenopus laevis* Daudin throughout metamorphosis. *Gen. comp. Endocrinol.*, **10**, 34–46.

Cooke, J. (1973). Properties of the primary organizational field in the embryo of *Xenopus laevis*. IV. Pattern formation and regulation following early inhibition of mitosis. *J. Embryol. exp. Morph.*, **30**, 49–62.

Coons, A.H. & Kaplan, M.H. (1950). Localization of antigen in tissue cells. *J. exp. Med.*, **91**, 1–14.

Cooper, E.L. (1967). Lymphomyeloid organs of Amphibia. I. Normal appearance during larval and adult stages of *Rana catesbeiana*. *J. Morph.*, **122**, 381–98.

Corner, M.A. (1963). Development of the brain of *Xenopus laevis* after removal of parts of the neural plate. *J. exp. Zool.*, **153**, 301–6.

Corner, M.A. (1964). Localization of capacities for functional development in the neural plate of *Xenopus laevis*. *J. comp. Neur.*, **123**, 243–56.

Crippa, M. & Gross, P.R. (1969). Maternal and embryonic contribution to the functional messenger-RNA of early development. *Proc. natn. Acad. Sci. U.S.A.*, **62**, 120–7.

Cronly-Dillon, J.R. & Muntz, W.R.A. (1965). The spectral sensitivity of the goldfish and the clawed toad tadpole under photopic conditions. *J. exp. Biol.*, **42**, 481–93.

Curry-Lindahl, K. (1956). Ecological studies on mammals, birds, reptiles and amphibians in the Eastern Belgian Congo. IV. Amphibians. *Ann. Mus. Roy. Congo. Belg. Ter., Sci. Zool.*, **42**, 53–65.

Curtis, A.S.G. (1957). The role of calcium in cell aggregation of *Xenopus* embryos. *Proc. R. Phys. Soc. Edinb.*, **26**, 25–32.

Curtis, A.S.G. (1958). A ribonucleoprotein from amphibian gastrulae. *Nature, Lond.*, **181**, 185.

Curtis, A.S.G. (1960a). Cortical grafting in *Xenopus laevis*. *J. Embryol. exp. Morph.*, **8**, 163–73.

Curtis, A.S.G. (1960b). Cell contact: Some physical considerations. *Amer. Nat.*, **94**, 37–56.

Curtis, A.S.G. (1961). Timing mechanisms in the specific adhesions of cells. *Expl. Cell Res. (suppl)*, **8**, 107–22.

Curtis, A.S.G. (1962). Morphogenetic interaction before gastrulation in the amphibian, *Xenopus laevis*. The cortical field. *J. Embryol. exp. Morph.*, **10**, 410–22.

Cuvier, G. (1829). *Le Règne Animal*, vol. 2 (2nd edn.). Paris: Deterville, Crochard.

Czołowska, R. (1969). Observations on the origin of the germinal cytoplasm in *Xenopus laevis*. *J. Embryol. exp. Morph.*, **22**, 229–51.

Daudin, F.M. (1803). *Histoire naturelle des rainettes, des grenouilles et des crapauds*. Bertrandet, Libraire Levrault, Paris, **13**, p. 92.

Davidson, E.H., Allfrey, V.G. & Mirsky, A.E. (1964). On the RNA synthesized during the lampbrush phase of amphibian oogenesis. *Proc. natn. Acad. Sci. U.S.A.*, **52**, 501–8.

Dawid, I.B. (1966). Evidence for the mitochondrial origin of frog egg cytoplasmic DNA. *Proc. natn. Acad. Sci. U.S.A.*, **56**, 269–76.

Dawid, I.B. & Wolstenholme, D.R. (1967). Ultracentrifuge and electron microscope studies on the structure of mitochondrial DNA. *J. molec. Biol.*, **28**, 233–45.

DeLaat, S.W., Luchtel, D. & Bluemink, J.G. (1973). The action of cytochalasin B during egg cleavage in *Xenopus laevis*: dependence on cell membrane permeability. *Devl. Biol.*, **31**, 163–77.

DeLanney, L.E. (1958). Influence of adult spleen on the development of embryos and larvae: an immune response? In: *The Chemical Basis of Development* (eds. W.D. McElroy & B. Glass), pp. 594–613. Baltimore: Johns Hopkins Press.

Denis, H. (1964). Effets de l'actinomycine sur le développement embryonnaire. I. Étude morphologique: suppression par l'actinomycine du compétence de l'ectoderme et du pouvoir inducteur de la lèvre blastoporale. *Devl. Biol.*, **9**, 435–57.

Dettelbach, H.R. (1952). Histostatic and cytostatic effects of some amino ketones upon tail regeneration in *Xenopus* larvae. *Revue Suisse Zool.*, **59**, 339–98.

Detwiler, S.R. (1936). *Neuroembryology: An Experimental Study*. New York & London: Haffner (Reprint) 1964.

Deuchar, E.M. (1956). Amino acids in developing tissues of *Xenopus laevis*. *J. Embryol. exp. Morph.*, **4**, 327–46.

Deuchar, E.M. (1957). Famous animals—8. The Axolotl. In: *New Biology*, **23**, 102–22 (eds. M.L. Johnson, M. Abercrombie & G.E. Fogg). London: Penguin Books Ltd.

Deuchar, E.M. (1958). Regional differences in catheptic activity in *Xenopus laevis* embryos. *J. Embryol. exp. Morph.*, **6**, 223–37.

Deuchar, E.M. (1960). The distribution of free leucine and its uptake into protein in *Xenopus laevis* embryos. In: *Symposium on Germ cells and Early Stages of Development* (ed. S. Ranzi), pp. 537–44. Milan: Fond. Baselli.

Deuchar, E.M. (1961). Amino-acid activation in tissues of early embryos. *Nature, Lond.*, **191**, 1006–7.

Deuchar, E.M. (1962). Amino-acid activation in embryonic tissues of *Xenopus laevis*. II. Hydroxamic acid formation in the presence of L-leucine. *Expl. Cell Res.*, **26**, 568–70.

Deuchar, E.M. (1963). Tracing amino acids from yolk protein into tissue protein. I. Incorporation of tritiated leucine into oocytes and its distribution in the early embryo of *Xenopus laevis*. *Acta Embryol. Morph. exp.*, **6**, 311–23.

Deuchar, E.M. (1967). Inducing activity in extracts from *Xenopus* embryos and from isolated dorsal lips. *J. Embryol. exp. Morph.*, **17**, 341–8.

Deuchar, E.M. (1970a). Effect of cell number on the type and stability of differentiation in amphibian ectoderm, *Expl. Cell Res.*, **59**, 341–3.

Deuchar, E.M. (1970b). Neural induction and differentiation with minimal numbers of cells. *Devl. Biol.*, **22**, 189–99.

Deuchar, E.M. (1971). Transfer of the primary induction stimulus by small numbers of amphibian ectoderm cells. *Acta Embryol. exp.*, pp. 387–96.

Deuchar, E.M. (1972). *Xenopus laevis* and developmental biology. *Biol. Rev.*, **47**, 37–112.

Deuchar, E.M. & Burgess, A.M.C. (1967). Somite segmentation in amphibian embryos: is there a transmitted control mechanism? *J. Embryol. exp. Morph.*, **17**, 349–59.

Deuchar, E.M., Weber, R. & Lehmann, F.E. (1957). Differential changes of catheptic activity in regenerating tails of *Xenopus* larvae, related to protein breakdown and total nitrogen. *Helv. Physiol. Acta*, **15**, 212–29.

Devilliers, C. (1924). On the anatomy of the breast-shoulder apparatus of *Xenopus*. *Ann. Transv. Mus.*, **10**, 197–211.

Devilliers, C. (1925). On the development of the 'epipubis' of *Xenopus*, *Ann. Transv. Mus.*, **12**, 129–35.

Dijkgraaf, S. (1962). The functioning and significance of the lateral-line organs. *Biol. Rev.*, **38**, 51–105.

Dodd, J.M. (1950). Ciliary feeding mechanisms in anuran larvae. *Nature, Lond.*, **165**, 283.

Dodd, J.M. (1973). Personal communication.

Dreyer, T.F. (1913). The 'Plathander' (*Xenopus laevis*). *Trans. R. Soc. S. Afr.*, **3**, 341–55.

Dreyer, T.F. (1914). The morphology of the tadpole of *Xenopus laevis*. *Trans. R. Soc. S. Afr.*, **4**, 241–8.

Driesch, H. (1910). Neue Versuche über die Entwicklung verschmolzener Echiniden-keime. *Wilhelm Roux Arch. EntwMech. Org.*, **30**, 8–23.

DuPasquier, L., Weiss, N. & Loor, F. (1972). Direct evidence for immunoglobulins on the surface of thymus lymphocytes of amphibian larvae. *Europ. Journ. Immunol.*, **2**, 366–70.

Eakin, R.M. & Lehmann, F.E. (1957). An electron microscope study of developing amphibian ectoderm. *Wilhelm Roux Arch. EntwMech. Org.*, **150**, 177–98.

Eeckhout, Y. (1969). Étude biochimique de la métamorphose caudale des amphibiens anoures. *Mem. Acad. r. Belg. Cl. Sci.*, **38**, 11–113.

Elek, S.D., Rees, T.A. & Gowing, N.F.C. (1962). Studies on the immune response in a poikilothermic species (*Xenopus laevis* Daudin). *Comp. Biochem. Physiol.*, **7**, 255–67.

Elkan, E. (1960). Some interesting pathological cases in amphibians. *Proc. Zool. Soc. Lond.*, **134**, 275–96.

Elkan, E. & Murray, R.W. (1951). New lateral line sensory organs in *Xenopus laevis* Daudin. *Nature, Lond.*, **168**, 477.

Elsdale, T.R., Fischberg, M. & Smith, S. (1958). A mutation that reduces nucleolar number in *Xenopus laevis*. *Expl. Cell Res.*, **14**, 642–3.

Elsdale, T.R., Gurdon, J.B. & Fischberg, M. (1960). A description of the technique for nuclear transplantation in *Xenopus laevis*. *J. Embryol. exp. Morph.*, **8**, 437–44.

Engels, H.G. (1935). Ueber Umbildungsvorgänge im Kardinalvenensystem bei Bildung der Urniere. *Morph. Jahrb.*, **76**, 345–74.

Esper, H. & Barr, H.J. (1964). A study of the developmental cytology of a mutation affecting nucleoli in *Xenopus* embryos. *Devl. Biol.*, **10**, 105–21.

Estensen, R.D., Rosenberg, M. & Sheridan, J.D. (1971). Cytochalasin B: micro-filaments and contractile processes. *Science, N.Y.*, **173**, 356–7.

Evans, M.J. (1969). Studies on the ribonucleic acid of early amphibian embryos. *Ph.D. Thesis*, University College, London.

Evennett, P.J. & Thornton, V.F. (1971). The distribution and development of gonadotropic activity in the pituitary of *Xenopus laevis*. *Gen. comp. Endocrinol.*, **16**, 606–7.

Ewer, D.W. (1959). A toad (*Xenopus laevis*) without haemoglobin. *Nature, Lond.*, **183**, 271.

Farquhar, M. & Palade, G.E. (1964). Functional organization of amphibian skin. *Proc. natn. Acad. Sci. U.S.A.*, **51**, 569–77.

Farquhar, M. & Palade, G.E. (1965). Cell junctions in amphibian skin. *J. Cell Biol.*, **26**, 263.

Faulhaber, I. (1972). Die Induktionsleistung subzellularer Fraktionen aus der Gastrula von *Xenopus laevis*. *Wilhelm Roux Arch. EntwMech. Org.*, **171**, 87–108.

Faulhaber, I. & Geithe, H.P. (1972). Nachweis deuterencephalspinocaudaler Induktionsfähigkeit in Gastrulaextrakten von *Xenopus laevis* nach Chromato-graphie am Hydroxyapatit. *Revue Suisse Zool.*, **79 (suppl).**, 103–17.

Ficq, A. (1961). Effets de l'actinomycine D et de la puromycine sur le métabolisme de l'oocyte en croissance. (Étude autoradiographique). *Expl. Cell Res.*, **34**, 581–93.

Firket, H. (1963). Effets des gonadotrophines placentaires sur les premiers stades de la spermatogénèse chez *Xenopus laevis*. *C. r. Séanc. Soc. Biol. Paris*, **157**, 1109–12.

Fischberg, M., Gurdon, J.B. & Elsdale, T.R. (1958). Nuclear transplantation in *Xenopus laevis*, *Nature, Lond.*, **181**, 424.

Fletcher, K. & Myant, N.B. (1960). Oxygen consumption of tadpoles during metamorphosis. *J. Physiol. Lond.*, **145**, 353–68.

Follett, B.K. & Heller, H. (1964). The neurohypophysial hormones of bony fishes. *J. Physiol. Lond.*, **172**, 92–106.

Foote, C.L. & Foote, F.M. (1960). Maintenance of gonads of *Xenopus laevis* in organ cultures. *Proc. Soc. exp. Biol. Med.*, **105**, 107–8.

Ford, P. (1971). Non-coordinate accumulation and synthesis of 5S RNA by ovaries of *Xenopus laevis*. *Nature, Lond.*, **233**, 561–4.

Ford, P.J. & Southern, E.M. (1973). Different sequences for 5S RNA in kidney cells and ovaries of *Xenopus laevis*. *Nature (New Biol.) Lond.*, **241**, 7–12.

Fox, E. & Irving, J.T. (1950). The effect of thyroid hormone on the ossification of the femur in *Xenopus laevis* tadpoles. *S. Afr. J. Med. Sci.*, **15**, 11–14.

Fox, H. (1963). The amphibian pronephros. *Q. Rev. Biol.*, **38**, 1–25.

Fox, H. (1970). Cilia in cloaca and hind gut of *Xenopus* larvae seen by electron microscopy. *Archs. Biol. Paris*, **81**, 1–20.

Fox, H. & Hamilton, L. (1964). Origin of the pronephric duct in *Xenopus laevis*. *Archs. Biol. Paris*, **75**, 245–51.

Fox, H. & Hamilton, L. (1971). Ultrastructure of diploid and haploid cells of *Xenopus laevis* larvae. *J. Embryol. exp. Morph.*, **26**, 81–98.

Fox, H., Mahoney, R. & Bailey, E. (1970). Aspects of the ultrastructure of the alimentary canal and associated glands of the *Xenopus laevis* larva. *Archs. Biol. Paris*, **81**, 21–50.

Fox, H. & Turner, S.C. (1967). A study of the relationship between the thyroid and larval growth in *Rana temporaria* and *Xenopus laevis*. *Archs. Biol. Paris*, **78**, 61–90.

Francis, E.T.B. (1934). *The Anatomy of the Salamander*. Oxford: Clarendon Press.

Freeman, G. (1963). Lens regeneration from the cornea in *Xenopus laevis*. *J. exp. Zool.*, **154**, 39–65.

Fried, P.H. & Rakoff, A.E. (1955). Inadequacies of the frog pregnancy test. *Obstet. Gynec. N.Y.*, **6**, 12–17.

Fruton, J.S., Hearn, W.H., Ingram, V.M., Wiggans, D.S. & Winitz, M. (1953). Synthesis of polymeric peptides in protein-catalysed transamidation reactions. *J. biol. Chem.*, **204**, 891–902.

Furshpan, E.J. & Potter, D.D. (1968). Low resistance junctions between cells in embryos and in tissue culture. *Curr. Top. devl. Biol.*, **3**, 95–128.

Gall, J.G. (1963). Chromosomes and cytodifferentiation. In: *Cytodifferentiation and Macromolecular Synthesis* (ed. M. Locke), pp. 119–43. New York: Academic Press.

Gallien, L. (1948). Réactivité du *Xenopus laevis* Daudin mâle aux gonadotrophines hypophysaires et chorioniques. Application au diagnostic biologique de la grossesse. *C. R. Acad Sci. Paris*, **226**, 1141–3.

Gallien, L. (1953). Inversion totale du sexe chez *Xenopus laevis* Daud. à la suite d'un traitement gynogène par le benzoate d'oestradiol. administré pendant la vie larvaire. *C. R. Acad. Sci. Paris*, **237**, 1565–6.

Gallien, L. (1955). Conséquences du changement de sexe pour la descendance d'un amphibien anoure, *Xenopus laevis*. *C. R. Acad. Sci Paris*, **241**, 998–1000.

Gallien, L. & Beetschen, J.-Cl. (1951). Extension et limites du pouvoir régénérateur des membres chez *Xenopus laevis* Daudin après la métamorphose. *C. r. Séanc. Soc. Biol.*, **145**, 874–6.

Gallien, L. & Foulgoc, M.C. (1960). Mise en évidence de steroïdes oestrogènes au cours du développement. *C. R. Acad. Sci. Paris*, **251**, 460–2.

Gasche, P. (1944). Beginn und Verlauf der Metamorphose bei *Xenopus laevis* Daud. Festlegung von Umwandlungsstadien. *Helv. physiol. Acta*, **2**, 607–26.

Gaze, R.M. (1970). *The Formation of Nerve Connections: a Consideration of Neuronal Specificity, Modulation and Comparable Phenomena*. New York and London: Academic Press.

Gaze, R.M., Jacobson, M. & Székely, G. (1965). On the formation of connexions by compound eyes in *Xenopus*. *J. Physiol. Lond.*, **176**, 409–17.

Gaze, R.M. & Keating, M.J. (1972). The visual system and 'neuronal specificity'. *Nature, Lond.*, **237**, 375–8.

Gitlin, G. (1939). Gravimetric studies of certain organs of *Xenopus laevis* (the South African Clawed Toad) under normal and experimental conditions. I. The oviduct. *S. Afr. J. med. Sci*, **4**, 41–52.

Gitlin, G. (1942). Effect of sheep anterior pituitary extract on the ovaries of *Xenopus laevis*. *S. Afr. J. med. Sci.*, **7**, 16–20.

Glaser, O.C. (1914). On the mechanism of the morphological differentiation of the nervous system. *Anat. Rec.*, **8**, 525–51.

Goodrich, E.S. (1938). *Structure and Development of Vertebrates*. London: Macmillan.

Goodwin, B. (1971). A model of early amphibian development. In: *Brit. Soc. exp. Biol. Symposium*, **25**, 417–28. (eds. D.D. Davies & M. Balls). Cambridge University Press.

Goos, H.J.Th. (1969). Hypothalamic neurosecretion and metamorphosis in *Xenopus laevis*. IV. The effect of extirpation of the presumed TRF cells and of a subsequent PTU treatment. *Z. Zellforsch. mikrosk. Anat.*, **97**, 449–58.

Graham, C.F., Arms, K. & Gurdon, J.B. (1966). The induction of DNA synthesis by frog egg cytoplasm. *Devl. Biol.*, **14**, 349–81.

Grant, P. (1953). Phosphate metabolism during oogenesis in *Rana temporaria*. *J. exp. Zool.*, **124**, 514–43.

Grant, P. & Wacaster, J.F. (1972). The amphibian gray crescent region—a site of developmental information? *Devl. Biol.*, **28**, 454–71.

Green, H., Goldberg, G.B., Schwartz, M. & Brown, D.D. (1968). The synthesis of collagen during the development of *Xenopus laevis*. *Devl. Biol.*, **18**, 391–400.

Grobbelaar, C.S. (1935). The musculature of the Pipid genus *Xenopus*. *S. Afr. J. Sci.*, **32**, 395.

Gross, J. (1964). Studies on the biology of connective tissues: the remodelling of collagen in metamorphosis. *Medicine*, **43**, 291–304.

Gurdon, J.B. (1959a). Tetraploid frogs. *J. exp. Zool.*, **141**, 519–38.

Gurdon, J.B. (1959b). The developmental capacity of nuclei taken from intestinal epithelium cells of feeding tadpoles. *J. Embryol. exp. Morph.*, **10**, 622–40.

Gurdon, J.B. (1960a). The transplantation of nuclei between *Xenopus laevis* and *X. l. victorianus*. In: *Symposium on germ cells and early stages of development* (ed. S. Ranzi), p. 482. Milan: Fond. Baselli.

Gurdon, J.B. (1960b). The effect of ultraviolet irradiation on uncleaved eggs of *Xenopus laevis*. *Q. Jl. microsc. Sci.*, **101**, 299–311.

Gurdon, J.B. (1964). The transplantation of living cell nuclei. *Adv. Morphogen.*, **4**, 1–43.

Gurdon, J.B. (1967). On the origin and persistence of a cytoplasmic state inducing nuclear DNA synthesis in frogs' eggs. *Proc. natn. Acad. Sci. U.S.A.*, **58**, 545–52.

Gurdon, J.B. (1968). Changes in somatic cell nuclei inserted into growing and maturing amphibian oocytes. *J. Embryol. exp. Morph.*, **20**, 401–14.

Gurdon, J.B., Birnstiel, M. & Speight, V.A. (1969). The replication of purified DNA introduced into living egg cytoplasm. *Biochim. biophys. Acta*, **174**, 614–28.

Gurdon, J.B. & Ford, P.J. (1970). Attachment of rapidly labelled RNA to polysomes in the absence of ribosomal RNA synthesis during normal cell differentiation. *Nature, Lond.*, **216**, 666–8.

Gurdon, J.B., Lane, C.D., Woodland, H.R. & Marbaix, G. (1971). Use of frog eggs and oocytes for the study of messenger RNA and its translation in living cells. *Nature, Lond.*, **233**, 177–82.

Gurdon, J.B. & Laskey, R.A. (1970). The transplantation of nuclei from single cultured cells into enucleate frogs' eggs. *J. Embryol. exp. Morph.*, **24**, 227–48.

Hadji-Azimi, I. (1971). Studies on *Xenopus laevis* immunoglobulins. *Immunology*, **21**, 463–74.

Hadley, M.E. & Quevedo, W.C. (1967). The role of epidermal melanocytes in adaptive color changes in amphibians. *Adv. Biol. Skin*, **8**, 337–59.

Haffner, K. von (1951). Die Pinealblase (Stirnorgan, Parietalorgan) von *Xenopus laevis* Daud. und ihre Entwicklung, Verlagerung und Degeneration. *Zool. Jahrb. (Abt. Anat.)*, **71**, 375–412.

Haglund, B. & Løvtrup, S. (1965). The influence of digitonin on water penetration in amphibian cells. *Expl. Cell Res.*, **37**, 200–6.

Hahn, H.P. von (1958). Die regionale Verteilung der Katalase-Aktivität im Schwanz von hungernden und gefütterten *Xenopus* larven. *Experientia*, **14**, 67–8.

Hallberg, R.L. & Brown, D.D. (1969). Coordinated synthesis of some ribosomal proteins and ribosomal RNA in embryos of *Xenopus laevis*. *J. molec. Biol.*, **46**, 393–411.

Hamburger, V. & Levi-Montalcini, R. (1949). Proliferation, differentiation and degeneration in the spinal ganglion of the chick embryo under normal and experimental conditions. *J. exp. Zool.*, **111**, 457–502.

Hamilton, L. (1963). An experimental analysis of the development of the haploid syndrome in embryos of *Xenopus laevis*. *J. Embryol. exp. Morph.*, **11**, 267–78.

Hamilton, L. (1969). The formation of somites in *Xenopus*. *J. Embryol. exp. Morph.*, **22**, 253–64.

Hamilton, L. & Tuft, P. (1972). The role of water-regulating mechanisms in the development of the haploid syndrome in *Xenopus laevis*. *J. Embryol. exp. Morph.*, **28**, 449–62.

Hanke, W. & Leist, K.H. (1971). The effect of ACTH and corticosteroids on carbohydrate metabolism during the metamorphosis of *Xenopus laevis*. *Gen. comp. Endocrinol.*, **16**, 85–96.

Hanke, W. & Neumann, U. (1972). Carbohydrate metabolism in amphibians. *Gen. comp. Endocrinol.*, **8, suppl. 3**, 198–208.

Hauser, R. (1965). Autonome Regenerationsleistungen des larvalen Schwanzes von *Xenopus laevis* und ihre Abhängigkeit von Zentralnervensystem. *Wilhelm Roux Arch. EntwMech. Org.*, **156**, 404–18.

Hauser, R. (1969). Abhängigkeit der normalen Schwanzenregeneration bei *Xenopus*-larven von einer diencephaler Faktor im Zentralkanal. *Wilhelm Roux Arch. EntwMech. Org.*, **163**, 221–47.

Hauser, R. (1972). Morphogenetic action of the subcommissural organ on tail regeneration in *Xenopus* larvae. *Wilhelm Roux Arch. EntwMech. Org.*, **169**, 170–84.

Hauser, R. & Lehmann, F.E. (1962). Regeneration in isolated tails of *Xenopus* larvae. *Experientia*, **18**, 83–4.

Hay, E.D. & Gurdon, J.B. (1967). Fine structure of the nucleolus in normal and mutant *Xenopus* embryos. *J. Cell Sci.*, **2**, 151–62.

Hayes, B.P. & Roberts, A. (1973). Synaptic junction development in the spinal cord of an amphibian embryo: an electron microscope study. *Z. Zellforsch.*, **137**, 251–69.

Heller, H. & Pickering, B.T. (1970). The distribution of vertebrate neurohypophyseal hormones and its relation to possible pathways for their evolution. *Internat. Encyclop. of Pharmacol. & Therapeut.*, **41** (1) chap. 3, 59–79.

Henderson, I.W., Edwards, B.R., Garland, H.O. & Chester Jones, I. (1972). Renal function in two toads, *Xenopus laevis* and *Bufo marinus*. *Gen. comp. Endocrinol. Suppl.*, **3**, 350–9.

Herner, A.E. & Frieden, E. (1960). Biochemistry of amphibian metamorphosis. VII. Changes in serum proteins during spontaneous and induced metamorphosis. *J. biol. Chem.*, **235**, 2845–51.

Hey, D. (1949). A report on the culture of the South African Clawed Frog *Xenopus laevis* (Daudin) at the Jonkershoek Inland Fish Hatchery. *Trans. R. Soc. S. Afr.*, **32**, 45–54.

Hobson, B.M. (1952). Conditions affecting the release of spermatozoa in male *Xenopus laevis* in response to chorionic gonadotrophin. *Q. Jl. exp. Physiol.*, **37**, 191–203.

Hobson, B.M. & Barr, W.A. (1965). The response of male Anura to follicle stimulating hormone (FSH) and to interstitial cell stimulating hormone (ICSH). *Gen. comp. Endocrinol.*, **5**, 686–8.

Hoff-Jørgensen, E. & Zeuthen, E. (1952). Evidence of cytoplasmic deoxyribosides in the frog's egg. *Nature, Lond.*, **169**, 245.

Hoffman, A.C. (1930). Opsomming van neuve navorsinge oor die opbau en ontogoenese van die zonalskelet by Amphibia, vernaanlik van *Cryptobranchus alleghaniensis, Necturus maculatus, Heleophryne* en *Xenopus laevis*. *S. Afr. J. Sci.*, **27**, 446–50.

Hogben, L. (1934). Spinal transections and the chromatic function in *Xenopus laevis*. *Trans. Roy. Soc. S. Africa*, **22**, xxvi-vii.

Hogben, L., Charles, E. & Slome, D. (1931). Studies on the pituitary. VIII. The relation of the pituitary gland to calcium metabolism and ovarian function in *Xenopus*. *J. exp. Biol.*, **8**, 345–54.

Hogben, L. & Gordon, C. (1930). Studies on the pituitary. VII. The separate identity of the pressor and melanophore principles. *J. exp. Biol.*, **7**, 286–92.

Holtfreter, J. (1933). Die totale Exogastrulation, eine Selbstablösung des Ektoderms vom Entomesoderm. Entwicklung und funktionelles Verhalten nervenlöser Organe. *Wilhelm Roux Arch. EntwMech. Org.*, **128**, 584–633.

Holtfreter, J. (1943). Experimental studies on the development of the amphibian pronephros. *Rev. Canad. Biol.*, **3**, 220–42.

Holtfreter, J. (1944). A study of the mechanics of gastrulation. *J. exp. Zool.*, **95**, 171–212.

Holtfreter, J. (1955). 1st Internat. Congr. Embryol. London (unpubl.).

Hope, J., Humphries, A.A. & Bourne, G.H. (1964a). Ultrastructural studies on developing oocytes of the salamander *Triturus viridescens*. I. The formation of yolk. *J. ultrastruct. Res.*, **10**, 547–56.

Hope, J., Humphries, A.A. & Bourne, G.H. (1964b). Ultrastructural studies on developing oocytes of the salamander *Triturus viridescens*. II. Early cytoplasmic changes and the formation of pigment. *J. ultrastruct. Res.*, **10**, 557–66.

Hörstadius, S. (1950). *The Neural Crest*. Oxford University Press.

Horton, J.D. (1969). Ontogeny of the immune response to skin allografts in relation to lymphoid organ development in the amphibian *Xenopus laevis* Daudin. *J. exp. Zool.*, **170**, 449–66.

Hughes, A.F.W. (1957). The development of the primary sensory system in *Xenopus laevis* (Daudin). *J. Anat.*, **91**, 323–8.

Hughes, 'A.F.W. (1959). Studies on embryonic and larval development in amphibia. II. The spinal motor-root. *J. Embryol. exp. Morph.*, **7**, 128–45.

Hughes, A.F.W. (1968). *Aspects of Neural Ontogeny*. New York & London: Logos/ Academic Press.

Hughes, A.F.W. & Lewis, P.R. (1961). Effect of limb ablation on neurones in *Xenopus* larvae. *Nature, Lond.*, **189**, 333–4.

Hughes, A.F.W. & Prestige, M.C. (1967). Development of behaviour in the hind limb of *Xenopus* larvae. *J. zool. Res.*, **152**, 347–59.

Hughes, A.F.W. & Tschumi, P.A. (1958). The factors controlling the development of the dorsal root ganglia and ventral horn in *Xenopus laevis* (Daudin). *J. Anat.*, **92**, 498–526.

Hughes, A.F.W. & Tschumi, P.A. (1960). Heterotopic grafting of the spinal cord in *Xenopus laevis* (Daudin). *Jl. R. microsc. Soc.*, **79**, 155–64.

Hunt, P.M. (1969). Effects of rotating neural tissue and underlying mesoderm in *Xenopus laevis* embryos. *Acta Embryol. exp.*, pp. 211–29.

Hutchison, J.B. & Poynton, J.C. (1963). A neurological study of the clasp reflex in *Xenopus laevis* (Daudin). *Behaviour*, **22**, 41–63.

Inoue, K. (1961). Serologically active groups of amphibian embryos. *J. Embryol. exp. Morph.*, **9**, 563–85.

Jacobson, C.O. (1968). Selective affinity as a working force in neurulation movements. *J. exp. Zool.*, **168**, 125–136.

Jacobson, C.O. (1970). Experiments on β-mercaptoethanol as an inhibitor of neurulation movements in amphibian larvae. *J. Embryol. exp. Morph.*, **23**, 463–71.

Jacobson, M. (1965). Development of neuronal specificity in retinal ganglion cells of *Xenopus*. *Devl. Biol.*, **17**, 202–18.

Jaeger, L.J. (1945). Glycogen utilization in the amphibian gastrula in relation to invagination and induction. *J. cell. comp. Physiol.*, **25**, 97–120.

Janssens, P.A. (1972). The influence of ammonia on the transition to ureotelism in *Xenopus laevis*. *J. exp. Zool.*, **182**, 357–66.

Jensen, P.K., Lehmann, F.E. & Weber, R. (1956). Catheptic activity in the regenerating tail of *Xenopus* larvae and its reaction to histostatic substances. *Helv. physiol. Acta*, **14**, 188–201.

Johnson, K.E. (1969). Altered contact behaviour of presumptive mesodermal cells from hybrid amphibian embryos arrested at gastrulation. *J. exp. Zool.*, **170**, 325–32.

Jolly, W.A. (1934). On recording lymph-heart beats. *Trans. R. Soc. S. Afr.*, **22**, v.

Jones, K.W. & Elsdale, T.R. (1963). The culture of small aggregates of amphibian embryonic cells *in vitro*. *J. Embryol. exp. Morph.*, **11**, 135–54.

Jones, T.R. (1938). The hand and foot musculature of three S. African anurans, *Rana fuscigula, Pyxicephalus adspersus* and *Xenopus laevis* and its evolutionary significance. *M. Sc. Thesis*, University of the Witwatersrand, Johannesburg.

Jurd, R.D. (1972). Personal communication.

Jurd, R.D. & McLean, N. (1970). An immunofluorescent study of the haemoglobins in metamorphosing *Xenopus*. *J. Embryol. exp. Morph.*, **23**, 299–309.

Kahn, J. (1962). The nucleolar organizer in the mitotic chromosome complement of *Xenopus laevis*. *Q. Jl. microsc. Sci.*, **103**, 407–9.

Kalk, M. (1960). Climate and breeding in *Xenopus laevis*. *S. Afr. J. Sci.*, **56**, 271–6.

Kalt, M.R. (1971). The relationship between cleavage and blastocoel formation in *Xenopus*. *J. Embryol. exp. Morph.*, **26**, 37–66.

Karfunkel, P. (1971). The role of microtubules and microfilaments in neurulation in *Xenopus*. *Devl. Biol.*, **25**, 30–56.

Keith, A. (1905). The nature of the mammalian diaphragm and pleural cavities. *J. Anat.*, **39**, 243–84.

Kelley, R.O. (1969). An electron microscope study of chordamesoderm-neurectoderm association in gastrulae of a toad, *Xenopus laevis*. *J. exp. Zool.*, **172**, 153–80.

Kelley, R.O., Nakai, G.S. & Guganig, M.E. (1971). A biochemical and ultrastructural study of RNA in yolk platelets of *Xenopus* gastrulae. *J. Embryol. exp. Morph.*, **26**, 181–93.

Kerr, T. (1965). Histology of the distal lobe of the pituitary of *Xenopus laevis* Daudin. *Gen. comp. Endocrinol.*, **5**, 232–40.

Kerr, T. (1966). The development of the pituitary in *Xenopus laevis* Daudin. *Gen. comp. Endocrinol.*, **6**, 303–11.

Kessel, R.G. (1969). Cytodifferentiation in the *Rana pipiens* oocyte. I. Association between mitochondria and nucleolus-like bodies in young oocytes. *J. ultrastruct. Res.*, **28**, 61–77.

Kezer, J. MacGregor, H.C. & Schabtach, E. (1971). Observations on the membranous components of amphibian oocyte nucleoli. *J. Cell Sci.*, **8**, 1–17.

Kidder, G.M., Ruben, L.N. & Stevens, J.M. (1973). Cytodynamics and ontogeny of the immune response of *Xenopus laevis* against sheep erythrocytes. *J. Embryol. exp. Morph.*, **29**, 73–85.

King, T. & Briggs, R. (1956). Serial transplantation of embryonic nuclei. *Cold Spring Harb. Symp. quant. Biol.*, **21**, 271–90.

Kobel, H.R., Brun, R.B. & Fischberg, M. (1973). Nuclear transplantation with melanophores, ciliated epidermal cells and the established cell line A-8 in *Xenopus laevis*. *J. Embryol. exp. Morph.*, **29**, 539–47.

Kocher-Becker, U. & Tiedemann, H. (1968). The inducing capacity of nucleic acids tested on amphibian ectoderm. *Wilhelm Roux Arch. EntwMech. Org.*, **160**, 375–400.

Kominck, H. (1961). Uber Herkunft, Bedeutung und Schicksal der Melanocyten im Cerebral-liquor von Krallenfroschlarven (*Xenopus laevis*). *Wilhelm Roux Arch. EntwMech. Org.*, **153**, 14–31.

Külemann, H. (1960). Untersuchungen der Pigmentbewegungen in embryonalen Melanophoren von *Xenopus laevis* in Gewebekulturen. *Zool. Jahrb. (Abt. Physiol. Zool.)*, **69**, 169–92.

Kunz, Y.W. (1973). Changes in lactate dehydrogenase isozyme pattern during the development of *Xenopus laevis* (Daudin). *Rev. Suisse Zool.*, **80**, 431–46.

Kunz, Y.W. & Hearn, J. (1967). Heterogeneity of lactate dehydrogenase in developing and adult *Xenopus laevis* Daud. *Experientia*, **23**, 683–6.

Landesman, R. & Gross, P.R. (1969). Patterns of macromolecular synthesis during development of *Xenopus laevis*. II. Identification of a 40S precursor to ribosomal RNA. *Devl. Biol.*, **19**, 244–60.

Landgrebe, F.W. (1939). The maintenance of reproductive activity in *Xenopus laevis* for pregnancy diagnosis. *J. exp. Biol.*, **16**, 89–95.

Landgrebe, F.W. & Waring, H. (1944). Biological assay and standardization of melanophore-expanding pituitary hormone. *Q. Jl. exp. Physiol.*, **33**, 1–18.

Laurent, R.F. (1972). *Exploration du Parc National des Virunga*, (2ᵉ série), Fasc. **22**, Amphibiens. 125 pp.

Leadley-Brown, A. (1970). *The African Clawed Toad*. London: Butterworth.

Lehman, H.E. (1953). Observations on macrophage behaviour in the fin of *Xenopus* larvae. *Biol. Bull. mar. biol. Lab., Woods Hole*, **105**, 490–5.

Lehmann, F.E. (1957). Die Schwanzenregeneration der *Xenopus*larve unter dem Einfluss phasenspezifischen Hemmstoffe. *Rev. Suisse Zool.*, **64**, 533–46.

Leslie, J.M. (1890). Notes on the habits and oviposition of *Xenopus laevis*. *Proc. zool. Soc. Lond.*, pp. 69–71.

Lohmann, K. (1972) Untersuchungen zur Frage der DNS-Konstanz in der Embryonalentwicklung. *Wilhelm Roux Arch. EntwMech. Org.*, **169**, 1–40.

Loveridge, A. (1925). Notes on East African batrachians, collected 1920–23, with the description of four new species. *Proc. zool. Soc. Lond.*, pp. 763–91.

Loveridge, A. (1932). New races of a skink (Siaphos) and a frog (*Xenopus*) from the Uganda Protectorate. *Proc. biol. Soc. Wash.*, **45**, 113–115.

Lykakis, J.J. (1969). The production of two molecular classes of antibody in the toad, *Xenopus laevis*, homologous with mammalian γM (19S) and γG (7S) immunoglobulins. *Immunology*, **16**, 91–8.

Lykakis, J.J. & Cox, F.E.G. (1968). Immunological responses of the toad *Xenopus laevis* to the antigens of the ciliate, *Tetrahymena pyriformis*. *Immunology*, **15**, 429–37.

McGarry, M.P. & Vanable, J.W. (1969a). The role of cell division in *Xenopus laevis* skin gland development. *Devl. Biol.*, **20**, 291–303.

McGarry, M.P. & Vanable, J.W. (1969b). The role of thyroxine in the formation of gland rudiments in the skin of *Xenopus laevis*. *Devl. Biol.*, **20**, 426–34.

MacGregor, H.C. (1965). The role of lampbrush chromosomes in the formation of nucleoli in amphibian oocytes. *Q. Jl. microsc. Sci.*, **106**, 215–28.

MacGregor, H.C. (1968). Nucleolar DNA in oocytes of *Xenopus laevis*. *J. Cell Sci.*, **3**, 437–44.

McLean, N. & Jurd, R.D. (1971). The haemoglobins of healthy and anaemic *Xenopus laevis*. *J. Cell Sci.*, **9**, 509–28.

MacMillan, G.J. (1972). The development of pigmentation in *Xenopus laevis* Daudin. *Ph.D. Thesis*, University of Glasgow.

MacMurdo, H.I. & Zalik, S.E. (1970). Embryonic cell surface: electrophoretic mobilities of blastula cells. *Experientia*, **26**, 406–8.

McMurray, V.M. (1954). The development of the optic lobes in *Xenopus laevis*: the effects of repeated crushing of the optic nerve. *J. exp. Zool.*, **125**, 247–63.

Malpoix, P., Quertier, J. & Brachet, J. (1963). The effects of β-mercaptoethanol on the morphogenetic movements of amphibian embryos. *J. Embryol. exp. Morph.*, **11**, 155–66.

Manning, M.J. (1971). The effect of early thymectomy on histogenesis of the lymphoid organs in *Xenopus laevis.*, *J. Embryol. exp. Morph.*, **26**, 219–29.

Manning, M.J. & Horton, J.D. (1969). Histogenesis of lymphoid organs in larvae of the South African Clawed Toad, *Xenopus laevis* (Daudin). *J. Embryol. exp. Morph.*, **22**, 265–77.

Marchalonis, J.J., Allen, R.B. & Saarni, E.S. (1970). Immunoglobulin classes of the clawed toad, *Xenopus laevis*. *Comp. Biochem. Physiol.*, **35**, 49–56.

Markert, C.L. & Whitt, G.S. (1968). Molecular varieties of isozymes. *Experientia*, **15**, 97–1008.

Marshall, R.A. & Nirenberg, M. (1969). RNA codons recognized by transfer-RNA from amphibian embryos and adults. *Devl. Biol.*, **19**, 1–11.

Masters, C.J. & Holmes, R.S. (1972). Isoenzymes and ontogeny. *Biol. Rev.*, **47**, 309–62.

Maynard-Smith, J. (1960). Continuous, quantized and modal variation. *Proc. R. Soc. B.*, **152**, 397–409.

Meban, C. (1973). The pneumonocytes in the lung of *Xenopus laevis*. *J. Anat.*, **114**, 235–44.

Michalowski, J. (1959). Über das Vorkommen von jugendlichen Hermaphroditismus bei *Xenopus laevis* Daud. *Biol. Zbl.*, **78**, 498–502.

Mikamo, K. & Witschi, E. (1964). Masculinization and breeding of the WW *Xenopus*. *Experientia*, **20**, 622–3.

Millard, N. (1941). The vascular anatomy of *Xenopus laevis* (Daudin). *Trans. R. Soc. S. Afr.*, **28**, 387–439.

Millard, N. (1945). The development of the arterial system of *Xenopus laevis*, including experiments on the destruction of the larval aortic arches. *Trans. R. Soc. S. Afr.*, **30**, 217–34.

Millard, N. (1949). The development of the venous system of *Xenopus laevis*. *Trans. R. Soc. S. Afr.*, **32**, 55–99.

Millard, N. & Robinson, J.T. (1955). *Dissection of the Spiny Dogfish and the Platanna*. Cape Town: Maskew Miller.

Miller, L. & Knowland, J. (1970). Reduction of ribosomal RNA synthesis and ribosomal RNA genes in a mutant of *Xenopus laevis* which organizes only a partial nucleolus. II. The number of ribosomal genes in animals of different nucleolar types. *J. molec. Biol.*, **53**, 329–38.

Milton, R.F. (1946). The *Xenopus* pregnancy test. *Br. med. J.*, **i**, 328.

Monnickendam, M.A. & Balls, M. (1973). Amphibian organ culture. *Experientia*, **15**, 1–17.

Mookerjee, J.S. & Das, S.K. (1939). Further investigations on the development of the vertebral column in Salientia (Anura). *J. Morph.*, **64**, 167–96.

Moore, J.A. (1960). Nuclear transfer of embryonic cells of the Amphibia. In: *Symposium on New Approaches in Cell Biology*, (ed. P.M.B. Walker), pp. 1–14. Cambridge University Press.

Moore, J.A. (1962). Nuclear transplantation and problems of specificity in developing embryos. *J. Cell Physiol.*, **60 (suppl. 1)**, 19–34.

Moulton, J.M., Jurand, A. & Fox, H. (1968). A cytological study of Mauthner's cells in *Xenopus laevis* and *Rana temporaria* during metamorphosis. *J. Embryol. exp. Morph.*, **19**, 415–31.

Munro, A.F. (1939). Nitrogen excretion and arginase activity during amphibian development. *Biochem. J.*, **33**, 1957–65.

Murray, R.W. (1956). The responses of the lateralis organs of *Xenopus laevis* to electrical stimulation by direct current. *J. Physiol. Lond.*, **134**, 408–20.

Needham, J. (1942). *Biochemistry and Morphogenesis*. Camb. Univ. Press.

Neuenschwander, P. (1972). Ultrastruktur und Jodaufnahme der Schilddrüse bei Larven des Krallenfrosches (*Xenopus laevis* Daud.). *Z. Zellforsch.*, **130**, 553–74.

Newth, D.R. (1949). The early development of the fore-limbs in *Xenopus laevis*. *Proc. zool. Soc. Lond.*, **118**, 559–67.

Newth, D.R. (1971). Personal communication.

Nicholas, J.S. & Hall, B.V. (1942). Experiments on developing rats. II. The development of isolated blastomeres and fused eggs. *J. exp. Zool.*, **90**, 441–59.

Nicolet, G. (1970). Is the presumptive notochord responsible for somite genesis in the chick? *J. Embryol. exp. Morph.*, **24**, 467–78.

Niehus, B. & Weber, R. (1961). Kennzeichnung und Verhalten der Mg^{++}—abhängigen Adenosine-triphosphatase im Schwanz der *Xenopus*larve während Wachstum und Metamorphose. *Helv. physiol. Acta*, **19**, 344–59.

Nieuwkoop, P.D. & Faber, J. (1956). *Normal Table of* Xenopus laevis *Daudin*. Amsterdam: N. Holland Publ. Co.

Nieuwkoop, P.D. & Florschütz, P.A. (1950). Quelques caractères spéciaux de la gastrulation et de la neurulation de l'oeuf de *Xenopus laevis* Daud. et de quelques autres anoures. *Archs. Biol. Paris*, **61**, 113–50.

Niu, M.-C. & Twitty, V. (1953). The differentiation of gastrula ectoderm in medium conditioned by axial mesoderm. *Proc. natn. Acad. Sci. U.S.A.*, **39**, 985–9.

Noble, G.K. (1931). *The Biology of the Amphibia.* New York & London: McGraw-Hill.

Nyholm, M., Saxèn, L., Toivönen, S. & Vainio, T. (1962). Electron microscopy of transfilter neural induction. *Expl. Cell Res.*, **28**, 209–12.

Okada, T.S. & Sirlin, J.L. (1960). The distribution of sulphur in the differentiating visceral cartilage of *Xenopus*. *J. Embryol. exp. Morph.*, **8**, 54–9.

Olsson, R. (1955). Structure and development of Reissner's fibre in the caudal end of *Amphioxus* and some lower vertebrates. *Acta zool. Stockh.*, **36**, 167–98.

Overton, J. (1963). Changes in cell fine structure during lens regeneration in *Xenopus laevis*. *J. Cell Biol.*, **24**, 211.

Overton, J. & Freeman, G. (1960). Lens regeneration in *Xenopus laevis*. *Anat. Rec.*, **137**, 386.

Palmer, J.F. & Slack, C. (1969). Effect of 'halothane' on electrical coupling in pre-gastrulation embryos of *Xenopus laevis*. *Nature, Lond.*, **223**, 1286–7.

Pantelouris, E.M., Knox, B. & Wallace, H. (1963). Iron in amphibian oocytes and embryos. *Expl. Cell Res.*, **32**, 469–75.

Parker, H.W. (1932). Scientific results of the Cambridge expedition to the East African lakes, 1930–1. 5. Reptiles and Amphibians. *J. Linn. Soc. Lond.*, **38**, 213–29.

Parker, H.W. (1936a). The amphibians of the Mamfe Division, Cameroons. I. Zoogeography and systematics. *Proc. zool. Soc. Lond.*, pp. 135–63.

Parker, H.W. (1936b). Reptiles and amphibians collected by the Lake Rudolf Rift Valley expedition, 1934. *Ann. Mag. Nat. Hist.*, **18**, 594–609.

Parker, F., Robbins, S.L. & Loveridge, A. (1947). Breeding, rearing and care of the S. African Clawed Frog, *Xenopus laevis*. *Amer. Nat.*, **81**, 38–49.

Paterson, N.F. (1939a). The head of *Xenopus laevis*. *Q. Jl. microsc. Sci.*, **81**, 161–272.

Paterson, N.F. (1939b). The olfactory organs and tentacles of *Xenopus laevis*. *S. Afr. J. Sci.*, **36**, 390–404.

Perkowska, E., MacGregor, H.C. & Birnstiel, M.L. (1968). Gene amplification in the oocyte nucleus of mutant and wild-type *Xenopus laevis*. *Nature, Lond.*, **217**, 649–50.

Perracca, M.G. (1898). Descrizione di una nuova specie di Amfibio del Gen. *Xenopus* Wagl. dell' Eritrea. *Boll. Mus. zool. Anat. comp. Torino*, **13**, no. 321, pp. 1–4.

Perrct, J.L. (1966). Les amphibiens du Cameroun. *Zool. Jahrb. (Syst.)*, **93**, 289–464.

Perriard, J.C. & Weber, R. (1971). Das Verhalten der ninhydrin-positive Stoffe im Schwanzgewebe von *Xenopus*-Larven während Wachstum und Metamorphose. *Wilhelm Roux Arch. EntwMech. Org.*, **166**, 365–76.

Pestell, R. (1971). Studies on the regulation of DNA synthesis in early animal development. *D. Phil. Thesis*, University of Oxford.

Pflugfelder, O. (1959). Atypische Gewebsdifferenzierung bei *Xenopus laevis* Daudin nach experimenteller Verhinderung der Metamorphose. *Wilhelm Roux Arch. EntwMech. Org.*, **151**, 229–41.

Piatt, J. (1942). Transplantation of aneurogenic forelimbs in Amblystoma punctatum. *J. exp. Zool.*, **91**, 79–101.

Politzer, W.M. (1963). Pregnancy diagnosis—haemagglutinin-inhibition method (prepuerin) compared with the *Xenopus laevis* test. *S. Afr. med. J.*, **37**, 905–10.

Politzer, W.M. & Simpson, F.E. (1963). Comparison of the *Xenopus laevis* pregnancy test with an haemagglutination inhibition (Prognosticon) test. *S. Afr. med. J.*, **37**, 931–2.

Poynton, J.C. (1964). Amphibia of Southern Africa. *Ann. Natal Mus.*, **17**, 1–334.

Prestige, M.C. (1965). Cell turnover in the spinal ganglia of *Xenopus laevis* tadpoles. *J. Embryol. exp. Morph.*, **13**, 63–72.

Prestige, M.C. (1967). The control of cell number in the lumbar ventral horns during the development of *Xenopus laevis* tadpoles. *J. Embryol. exp. Morph.*, **18**, 359–87.

Ranzi, S. & Gavarosi, G. (1959). Dimensions of the notochord and somites in embryos of *Xenopus laevis* treated with thiocyanate. *J. Embryol. exp. Morph.*, **7**, 117–21.

Redshaw, M.R. & Nicholls, T.J. (1971). Oestrogen biosynthesis by ovarian tissue of the South African Clawed Toad, *Xenopus laevis* Daudin. *Gen. comp. Endocrinol.*, **16**, 85–96.

Reed, S.C. & Stanley, H.P. (1972). Fine structure of spermatogenesis in the South African Clawed Toad, *Xenopus laevis* Daudin. *J. Ultrastruct. Res.*, **41**, 277–95.

Rich, K. & Weber, R. (1968). Die Metamorphosereaktion bei Xenopuslarven nach kurzfristiger Thyroxinbehandlung. *Rev. Suisse Zool.*, **75**, 650–60.

Rimer, G.E.G. (1931). Histological and morphological study of the endocrine glands of *Xenopus laevis*. *Ph.D. Thesis*, Univ. of Cape Town.

Robbins, S.L., Parker, F. & Bianco, P.D. (1947). The reaction of the male South African Clawed Frog (*Xenopus laevis*) to gonadotropins. *Endocrinology*, **40**, 227–9.

Roberts, A. & Stirling, C.A. (1971). The properties and propagation of a cardiac-like impulse in the skin of young tadpoles. *Z. vergl. Physiol.*, **71**, 295–310.

Rose, W. (1962). *The Reptiles and Amphibians of Southern Africa*. Cape Town: Maskew Miller (2nd. edition).

Rose, W. & Hewitt, J. (1927). Description of a new species of *Xenopus* from the Cape Peninsula. *Trans. R. Soc. S. Afr.*, **14**, 343–6.

Rostand, J. (1951). Parthenogénèse expérimentale chez le xénope (*Xenopus laevis*). *C. r. Séanc. Soc. Biol.*, **145**, 1453–4.

Roux, W. (1888). Über die künstliche Hervorbringung 'halber' Embryonen durch Zerstörung einer der beiden ersten Furchungszellen, sowie über die Nachentwicklung der fehlenden Korperhälfte. *Virchows Arch.*, **114**, 22–60.

Ruben, L.N., Stevens, S.J.M. & Kidder, G.M. (1972). Suppression of the allograft response by implants of mature lymphoid tissues in larval *Xenopus laevis*. *J. Morph.*, **138**, 457.

Ruddle, F.H. & Harrington, L. (1967). Tissue-specific esterase isozymes of the mouse (*Mus musculus*). *J. exp. Zool.*, **166**, 51–64.

Russell, I.J. (1971a). The role of the lateral-line efferent system in *Xenopus laevis*. *J. exp. Biol.*, **54**, 621–42.

Russell, I.J. (1971b). The pharmacology of efferent synapses in the lateral-line system of *Xenopus laevis*. *J. exp. Biol.*, **54**, 643–58.

Ryffel, G., Hagenbüchle, O. & Weber, R. (1973). Unchanged number of rRNA genes in liver and tail muscle of *Xenopus* larvae during thyroxin induced metamorphosis. *Cell Differentiation*, **2**, 191–8.

Ryffel, G. & Weber, R. (1973). Changes in the pattern of RNA synthesis in different tissues of *Xenopus* larvae during induced metamorphosis. *Expl. Cell Res.*, **77**, 79–88.

Sanderson, I.T. (1936). The amphibians of the Mamfe Division, Cameroons. II. Ecology of the frogs. *Proc. zool. Soc. Lond.*, pp. 165–208.

Savage, R.M. (1965). External stimulus for the natural spawning of *Xenopus laevis*. *Nature, Lond.*, **205**, 618–9.

Saxèn, L. (1960). Secretory changes in the mucoid cells of the anterior pituitary. *Acta path. microbiol. Scand.*, **48**, 341–50.

Saxèn, L. & Toivönen, S. (1955). The development of the ultimobranchial body in *Xenopus laevis* Daudin and its relation to the thyroid gland and epithelial bodies. *J. Embryol. exp. Morph.*, **31**, 376–84.

Saxèn, L. & Toivönen, S. (1962). *Primary Embryonic Induction.* New York & London: Logos/Academic Press.

Saxèn, L., Saxèn, E., Toivönen, S. & Salimaki, K. (1957a). Quantitative investigations into the anterior pituitary-thyroid mechanism during frog metamorphosis. *Endocrinology*, **61**, 35–44.

Saxèn, L., Saxèn, E., Toivönen, S. & Salimaki, K. (1957b). The anterior pituitary and the thyroid function during normal and abnormal development of the frog. *Soumal. eläin-ja kasvit. Seur. van Julk.*, **18**, 1–44.

Scheer, U. (1973). Nuclear pore flow rate of ribosomal RNA and chain growth rate of its precursor during oogenesis of *Xenopus laevis. Devl. Biol.*, **30**, 13–28.

Schmidt, K.P. & Inger, R.F. (1959). Amphibians, exclusive of the genera *Afrixalus* and *Hyperolius. Explor. Parc Natn. Umpeba, Miss. G.F. De Witte*, **56**, 1–264.

Schoepfle, G.M., Atkins, E. & Schafer, L.A. (1965). Iodoacetate depression in *Xenopus* sciatic single nerve fibres. *Am. J. Physiol.*, **208**, 720–3.

Schrire, V. (1939). Changes in plasma inorganic phosphate associated with endocrine activity in *Xenopus laevis.* II. The effect of insulin on hypophysectomized animals. *S. Afr. J. med. Sci.*, **4 (suppl.)**, 1–3.

Schroeder, T.E. (1970). Neurulation in *Xenopus laevis.* An analysis and model based upon light and electron microscopy. *J. Embryol. exp. Morph.*, **23**, 427–62.

Schultheiss, H., Hanke, W. & Maetz, J. (1972). Hormonal regulation of the skin diffusional permeability to water during development and metamorphosis of *Xenopus laevis* Daudin. *Gen. comp. Endocrinol.*, **18**, 400–3.

Seidel, F. (1960). Die Entwicklungsfähigkeiten isolierter Furchungzellen aus dem Ei des Kaninchens *Oryctolagus cuniculus. Wilhelm Roux Arch. EntwMech. Org.*, **152**, 43–130.

Sclman, G.G. (1958). An ultra-violet light method for producing haploid amphibian embryos. *J. Embryol. exp. Morph.*, **6**, 634–7.

Selman, G.G. & Pawsey, G.J. (1965). The utilization of yolk platelets by tissues of *Xenopus* embryos studied by a safranin staining method. *J. Embryol. exp. Morph.*, **14**, 191–212.

Selman, G.G. & Perry, M.M. (1970). Ultrastructural changes in the surface layers of the newt's egg in relation to the mechanism of its cleavage. *J. Cell Sci.*, **6**, 207–27.

Shaffer, B.M. (1963). The isolated *Xenopus laevis* tail: a preparation for studying the central nervous system and metamorphosis in culture. *J. Embryol. exp. Morph.*, **11**, 77–90.

Shapiro, B.G. (1933). The topography and histology of the parathyroid glandules in *Xenopus laevis. J. Anat.*, **68**, 39–44.

Shapiro, B.G. & Zwarenstein, H. (1934). The effect of hypophysectomy and castration on muscle creatine in *Xenopus laevis. Trans. R. Soc. S. Afr.*, **23**, xvi.

Shapiro, H.A. (1937). The biological basis of sexual behaviour in amphibians. IV. Influence of the injection of anterior pituitary extracts on the mating reflex of hypophysectomized and of gonadectomized South African Clawed Toads. *J. exp. Biol.*, **14**, 38–47.

Shapiro, H.A. (1939). Ovulation in *Xenopus* induced by certain steroids. *S. Afr. J. med. Sci.*, **4**, 21–31.

Shelton, P.M.J. (1970). The lateral line system at metamorphosis in *Xenopus laevis* (Daudin). *J. Embryol. exp. Morph.*, **24**, 511–24.

Sims, R.T. (1962). Transection of the spinal cord in developing *Xenopus laevis. J. Embryol. exp. Morph.* **10**, 115–26.

Singer, M. (1952). The influence of the nerve in regeneration of the amphibian extremity. *Q. Rev. Biol.*, **27**, 169–200.

Sirlin, J.L. (1956). Tracing morphogenetic movements by means of labelled cells. *Wilhelm Roux Arch. EntwMech. Org.*, **148**, 489–93.

Skowron, S., Jordan, M. & Rogulski, H. (1956). Regenerative capacity of tadpoles inhibited in growth and development. *Nature, Lond.*, **178**, 602–3.

Slack, C. & Palmer, J.F. (1969). The permeability of intercellular junctions in early embryos of *Xenopus laevis*, studied with fluorescent tracer. *Expl. Cell Res.*, **55**, 416–9.

Slome, D. & Hogben, L. (1929). The time factor in chromatic responses of *Xenopus laevis*. *Trans. R. Soc. S. Afr.*, **17**, 141–50.

Slome, D. & Hogben, L. (1934). Further observations on the relation of the pituitary gland to the chromatic function of *Xenopus laevis*. *Trans. R. Soc. S. Afr.*, **22**, xvi.

Smit, A.L. (1953). The ontogenesis of the vertebral column of *Xenopus laevis* (Daudin), with special reference to the segmentation of the metotic region of the skull. *Annale Univ. Stellenbosch*, **29A**, 79–136.

Smith, L.D. & Ecker, R.T. (1969). Role of the oocyte nucleus in physiological maturation in *Rana pipiens*. *Devl. Biol.*, **19**, 281–308.

Smith, S. (1958). Induction of triploidy in the South African Clawed Frog, *Xenopus laevis* (Daudin). *Nature, Lond.*, **181**, 290.

Spemann, H. (1901). Entwicklungsphysiologische Studien am Tritonei. *Wilhelm Roux Arch. EntwMech. Org.*, **12**, 224–64.

Spemann, H. (1903). Uber die Linsenbildung bei defekter Augenblase. *Anat. Anz.*, **23**, 457–64.

Spemann, H. (1924). Uber Organisatoren in der tierischen Entwicklung. *Naturwiss.*, **12**, 1092–4.

Spemann, H. (1938). *Embryonic Development and Induction*. New York: Haffner reprint, 1962.

Sperry, R.W. (1965). Embryogenesis of behavioural nerve nets. In: *Organogenesis*. (eds. R.L. DeHaan & H. Ursprung), pp. 161–86. New York & London: Holt, Reinhart & Winston.

Spofford, W.R. (1948). Observations on the posterior part of the neural plate in *Amblystoma*. *J. exp. Zool.*, **107**, 123–59.

Spratt, N.T. (1957). Analysis of the organizer center in the early chick embryo. III. Regulative properties of the chorda and somite centers. *J. exp. Zool.*, **135**, 319–54.

Stanisstreet, M. (1972). An immunochemical study of the proteins of the oocytes and early embryos of the South African Clawed Toad, *Xenopus laevis*. Ph.D. Thesis, University of Bristol.

Stanisstreet, M. & Deuchar, E.M. (1972). Appearance of antigenic material in gastrula ectoderm after neural induction. *Cell Diffn.*, **1**, 15–18.

Stebbins, R.C. & Eakin, R.M. (1958). The role of the 'third eye' in reptilian behaviour. *Am. Mus. Novit.*, no. 1870, pp. 1–40.

Steinman, R.M. (1968). An electron microscope study of ciliogenesis in developing epidermis and trachea in the embryo of *Xenopus laevis*. *Am. J. Anat.*, **122**, 19–56.

Sterba, G. (1950). Mitteilungen über die Altersinvolution des Amphibienthymus. I. Volumetrisch Bestimmungen am Thymus des Krallenfrosches. *Anat. Anz.*, **99**, 106–14.

Stevens, L.C. (1954). The origin and development of chromatophores of *Xenopus laevis* and other anurans. *J. exp. Zool.*, **125**, 221–46.

Stone, L.S. (1954). Further experiments on lens regeneration in eyes of the adult newt, *Triturus viridescens*. *Anat. Rec.*, **120**, 599–624.

Straznicky, K. (1973). The formation of the optic fibre projection after partial tectal removal in *Xenopus. J. Embryol. exp. Morph.*, **29**, 397–409.

Straznicky, K. & Gaze, R.M. (1971). The growth of the retina in *Xenopus laevis*: an autoradiographic study. *J. Embryol. exp. Morph.*, **26**, 67–79.

Straznicky, K. & Gaze, R.M. (1972). The development of the tectum in *Xenopus laevis*: an autoradiographic study. *J. Embryol. exp. Morph.*, **28**, 87–115.

Straznicky, K., Gaze, R.M. & Keating, M.J. (1971). The retinotectal projections after uncrossing the optic chiasma in *Xenopus* with one compound eye. *J. Embryol. exp. Morph.*, **26**, 523–42.

Sudarawati, S. & Nieuwkoop, P.D. (1971). Mesoderm formation in the anuran *Xenopus laevis* (Daudin). *Wilhelm Roux Arch. EntwMech. Org.*, **166**, 189–204.

Swanson, R.F. & Dawid, I.B. (1970). The mitochondrial ribosome of *Xenopus laevis*. *Proc. natn. Acad. Sci. U.S.A.*, **66**, 117–24.

Sze, L.C. (1953). Changes in the amount of desoxyribonucleic acid in the development of *Rana pipiens. J. exp. Zool.*, **122**, 577–601.

Takayanagi, N. (1958). Immunochemical analysis of differentiation. *J. Juzen Med. Soc.* (Japan), **60**, 701–51.

Tarin, D. (1971a). Scanning electron microscopical studies of the embryonic surface during gastrulation and neurulation in *Xenopus laevis. J. Anat.*, **109**, 535–47.

Tarin, D. (1971b). Histological features of neural induction in *Xenopus laevis*. *J. Embryol. exp. Morph.*, **26**, 543–70.

Tarin, D. (1973). Histochemical and enzyme digestion studies on neural induction in *Xenopus laevis. Differentiation*, **1**, 109–26.

Tarin, D. & Sturdee, A.P. (1971). Early limb development of *Xenopus laevis. J. Embryol. exp. Morph.*, **26**, 169–79.

Tarin, D. & Sturdee, A. (1973). Histochemical features of hind limb development in *Xenopus laevis. J. Anat.*, **114**, 101–7.

Tarkowski, A.K. & Wroblewska, J. (1967). Development of blastomeres of mouse eggs isolated at the 4- and 8-cell stage. *J. Embryol. exp. Morph.*, **18**, 155–80.

Tata, J.A. (1965). Turnover of nuclear and cytoplasmic RNA at the onset of induced amphibian metamorphosis. *Nature, Lond.*, **207**, 378–81.

Tencer, R. (1961). The effect of 5-flourodeoxyuridine on amphibian embryos. *Expl. Cell Res.*, **23**, 418–9.

Thomas, N. & Deuchar, E.M. (1971). Synthesis of high-molecular-weight RNA in *Xenopus* ectoderm after neural induction. *Acta Embryol. exp.*, pp. 195–200.

Thornton, V.F. (1971a). A bioassay for progesterone and gonadotrophins based on the meiotic division of *Xenopus* oocytes *in vitro. Gen. comp. Endocrinol.*, **16**, 599–605.

Thornton, V.F. (1971b). The effect of change of background colour on the MSH content of the pituitary in *Xenopus laevis. Gen. comp. Endocrinol.*, **17**, 554–60.

Tiedemann, H. (1967). Biochemical aspects of primary induction and determination. In: *The Biochemistry of Animal Development*, (ed. R. Weber), vol II., pp. 4–56. New York: Academic Press.

Tinsley, R.C. (1973). Personal communication.

Tinsley, R.C. (1973). Studies on the ecology and systematics of a new species of clawed toad, the genus *Xenopus*, from western Uganda. *J. Zool. Lond.*, **169**, 1–27.

Tinsley, R.C. (1974). The morphology and distribution of *Xenopus vestitus* Laurent, 1972 (Anura: Pipidae) from Central Africa. *J. Zool. Lond.* (in press).

Toivönen, S. (1952). Insufficiency of light as a cause of neoteny in *Xenopus laevis*. *Acta endocr. Copenh.*, **10**, 243–54.

Townes, P.L. & Holtfreter, J. (1955). Directed movements and selective adhesion of embryonic cells. *J. exp. Zool.*, **128**, 53–150.

Tschumi, P.A. (1956). Die Bedeutung der Epidermisleiste für die Entwicklung der Beine von *Xenopus laevis* Daud. *Rev. Suisse Zool.*, **63**, 707–16.

Tschumi, P.A. (1957). The growth of the hindlimb bud of *Xenopus* and its dependence upon the epidermis. *J. Anat.*, **91**, 243–54.

Tuft, P. (1961). Role of water-regulating mechanisms in amphibian morphogenesis: a quantitative hypothesis. *Nature, Lond.*, **192**, 1049–51.

Tuft, P. (1965). The uptake and distribution of water in the developing amphibian embryo. *Brit. Soc. exp. Biol. Symposium*, **19**, 385–402.

Turing, A.M. (1952). The chemical basis of morphogenesis. *Phil. Trans. Roy. Soc. B*, **237**, 37–72.

Turner, R.J. (1969). The functional development of the reticulo-endothelial system in the toad, *Xenopus laevis* (Daudin). *J. exp. Zool.*, **170**, 467–80.

Turner, R.J. (1973). Response of the toad, *Xenopus laevis* to circulating antigens. II. Responses after splenectomy. *J. exp. Zool.*, **183**, 35–46.

Turner, R.J. & Manning, M.J. (1973). Responses of the toad, *Xenopus laevis*, to circulating antigens. I. Cellular changes in the spleen. *J. exp. Zool.*, **183**, 21–34.

Vainio. T., Saxèn, L. & Toivönen, S. (1960). Transfer of the antigenicity of guinea pig bone marrow implants to the graft tissue in explantation experiments. *Experientia*, **16**, 27–9.

Van Gansen, P. (1967). Étude au microscope électronique des structures ribosomales du cytoplasme au cours de la ségmentation de l'œuf de *Xenopus laevis*. *Expl. Cell Res.*, **47**, 157–66.

Van Gansen, P. & Schram, A. (1968). Ultrastructure et cytochimie ultrastructurale de la vésicule germinative et du cytoplasme perinucléaire de l'oocyte mûr de *Xenopus laevis*. *J. Embryol. exp. Morph.*, **20**, 375–89.

Van Gansen, P. & Schram, A. (1972). Evolution of the nucleoli during oogenesis in *Xenopus laevis* studied by electron microscopy. *J. Cell Sci.*, **10**, 339–67.

Van Pletzen, R. (1953). Ontogenesis and morphogenesis of the breast–shoulder apparatus of *Xenopus laevis*. *Annale Univ. Stellenbosch*, **29A**, 137–84.

Vanable, J.W. (1964). Granular gland development during *Xenopus laevis* metamorphosis. *Devl. Biol.*, **10**, 331–57.

Vanable, J.W. & Mortensen, R.D. (1966). Development of *Xenopus laevis* skin glands in organ culture. *Expl. Cell Res.*, **44**, 436–42.

Veerdonk, F.C.G. van de (1960). Serotonin, a melanocyte-stimulating component in the dorsal skin secretion of *Xenopus laevis*. *Nature, Lond.*, **184**, 948–9.

Vilimikova, V. & Nedvídek, J. (1962). Changes in DNA content during early embryonic development of *Xenopus laevis* (Daudin) and *Rana temporaria* (L.). *Folia Biol. Krakow*, **8**, 381–9.

Vilter, V. (1946). Interversion dorso-ventrale des champs rétiniens chez le *Xenopus laevis* et répercussions sur le réflexe pigmentaire rétino-cutané. *C. r. Séanc. Soc. Biol.*, **140**, 760–3.

Volpe, E.P. & Gebhardt, B.M. (1965). Effect of dosage on the survival of embryonic homotransplants in the leopard frog, *Rana pipiens*. *J. exp. Zool.*, **160**, 11–28.

Waddington, C.H. (1952). *Principles of Embryology*. London: Allen & Unwin.

Waddington, C.H. (1962). *New Patterns in Genetics and Development*. New York and London: Columbia University Press.

Waddington, C.H. & Deuchar, E.M. (1953). Studies on the mechanism of meristic segmentation. I. The dimensions of somites. *J. Embryol. exp. Morph.*, **1**, 349–56.

Waddington, C.H. & Perkowska, E. (1965). Synthesis of ribonucleic acid by different regions of the early amphibian embryo. *Nature, Lond.*, **207**, 1244–6.

Waddington, C.H. & Sirlin, J.L. (1964). The incorporation of labelled amino-acids into amphibian embryos. *J. Embryol. exp. Morph.*, **2**, 340–7.

Wager, V.A. (1965). The Platanna *Xenopus laevis*. *Afr. Wild Life*, **9**, 49–53.

Wagler, H. (1827). Footnote to letter of H. Boie in *Isis*, **20**, 726.

Waldvogel, H. (1965). Untersuchungen über den Zellstoffwechsel früher Onto-genesestadien von *Xenopus laevis* (Daudin), mit besonderer Berücksichtigung der Funktion und Verteilung des Glykogens. *Wilhelm Roux Arch. EntwMech. Org.*, **156**, 20–48.

Wallace, H. & Birnstiel, M.L. (1966). Ribosomal cistrons and the nucleolar organizer. *Biochim. biophys. Acta*, **114**, 296–310.

Wallace, H., Morray, J. & Langridge, W.H.R. (1971). Alternative model for gene amplification. *Nature, Lond. (New Biol.)*, **230**, 201–3.

Wallace, R.A. & Dumont, J.N. (1968). The induced synthesis and transport of yolk proteins and their accumulation by the oocyte in *Xenopus laevis*. *J. cell. comp. Physiol.*, **72 (suppl. 1)**, 73–89.

Wallace, R.A. & Jared, D.W. (1969). Studies on amphibian yolk. VIII. The oestrogen-induced hepatic synthesis of a serum lipophosphoprotein and its selective uptake by the ovary and transformation into yolk platelet proteins in *Xenopus laevis*. *Devl. Biol.*, **19**, 498–526.

Ward, R.J. (1962). The origin of protein and fatty yolk in *Rana pipiens*. 2. Electron microscopical and cytochemical observations of young and mature oocytes. *J. Cell Biol.*, **14**, 309–41.

Waring, H., Landgrebe, F.W. & Neill, H. (1941). Ovulation and oviposition in the Anura. *J. exp. Biol.*, **18**, 11–25.

Wartenberg, H. (1964). Experimentelle Untersuchungen über die Stoffaufnahme durch Pinocytose während der Vitellogenese des Amphibienoocyten. *Z. mikrosk. anat. Forsch.*, **63**, 1004–10.

Wassersug, R. (1972). The mechanism of ultraplanktonic entrapment in anuran larvae. *J. Morph.*, **137**, 279–88.

Weber, R. (1957). Die Kathepsinaktivität im Schwanz von *Xenopus laevis* während Wachstum und Metamorphose. *Rev. Suisse Zool.*, **64**, 326–36.

Weber, R. (1962). Induced metamorphosis in isolated tails of *Xenopus laevis*. *Experientia*, **18**, 84–5.

Weber, R. (1964). Ultrastructural changes in regressing tail muscles of *Xenopus* larvae at metamorphosis. *J. Cell Biol.*, **22**, 481–7.

Weber, R. (1965). Inhibitory effects of actinomycin D on tail atrophy in *Xenopus* larvae at metamorphosis. *Experientia*, **21**, 665–6.

Weber, R. (1967). Biochemistry of amphibian metamorphosis. In: *The Biochemistry of Animal Development* (ed. R. Weber), vol. II, pp. 227–301. New York: Academic Press.

Weber, R. & Boell, E.J. (1955). Uber die Cytochromeoxydaseaktivität der Mito-chondrien von frühen Entwicklungsstadien des Krallenfrosches (*Xenopus laevis* Daud.). *Rev. Suisse Zool.*, **62**, 260–8.

Weber, R. & Boell, E.J. (1962). Enzyme patterns in isolated mitochondria from embryonic and larval tissues of *Xenopus*. *Devl. Biol.*, **4**, 452–72.

Weber, R. & Niehus, B. (1961). Zur Aktivität der Sauren Phosphatase im Schwanz der *Xenopus*-larven während Wachstum und Metamorphose. *Helv. physiol. Acta*, **19**, 103–17.

Weinbrenn, C. (1925). Some endoparasites found in certain Amphibia in the Transvaal. *M.Sc. Thesis*, Univ. of Witwatersrand, Johannesburg.

Weisman, A.I. & Coates, C.W. (1944a). *The South African frog, (Xenopus laevis) in pregnancy diagnosis: a Research Bulletin*. New York: Biologic Research Foundation.

Weisman, A.I. & Coates, C.W. (1944b). Effect of illumination on the egg-extrusion reaction of *Xenopus laevis* in the frog test for pregnancy. *Endocrinology*, **35**, 68–9.

Weiss, P. & Rosetti, F. (1957). Growth responses of the opposite sign among different neuron types exposed to thyroid hormone. *Proc. natn. Acad. Sci. U.S.A.*, **37**, 540–56.

Weissmann, G. (1961). Changes in connective tissue and intestine caused by vitamin A in Amphibia, and their acceleration by hydrocortisone. *J. exp. Med.*, **114**, 581–92.

Weisz, P.B. (1945a). The development and morphology of the larva of the South African Clawed Toad, *Xenopus laevis*. I. The third-form tadpole. *J. Morph.*, **77**, 163–91.

Weisz, P.B. (1945b). The development and morphology of the larva of the South African Clawed Toad, *Xenopus laevis*. II. The hatching and the first- and second-form tadpoles. *J. Morph.*, **77**, 193–217.

Whitehouse, R.H. & Grove, A.J. (1947). *Dissection of the Frog*. Cambridge: University Tutorial Press.

Wickbom, T. (1945). Cytological studies on Dipnoi, Urodela, Anura and *Emys*. *Hereditas*, **31**, 241–346.

Witschi, E. (1938). Studies on sex differentiation and sex determination in amphibians. V. Range of the cortex-medulla antagonism in parabiotic twins of Ranidae and Hylidae. *J. exp. Zool.*, **78**, 113–45.

Witschi, E. (1955). The bronchial columella of the ear of larval Ranidae. *J. Morph.*, **96**, 497–511.

Woerdeman, M.W. (1933). Uber den Glykogenstoffwechsel des Organisationszentrums in der Amphibiengastrula. *Proc. K. ned. Akad. Wet.*, **36**, 477–81.

Wolpert, L., Hicklin, J. & Hornbruch, A. (1971). Positional information and pattern regulation in regeneration of Hydra. *Brit. Soc. exp. Biol. Symposium*, **25**, 417–28. (eds. D.D. Davies & M. Balls).

Woodland, H.R. (1969). The phosphorylation of thymidine by oocytes and eggs of *Xenopus laevis* Daudin. *Biochim. biophys. Acta*, **186**, 1–12.

Woodland, H.R. & Gurdon, J.B. (1968a). The relative rates of synthesis of DNA, sRNA and rRNA in the endodermal region and other parts of *Xenopus laevis* embryos. *J. Embryol. exp. Morph.*, **19**, 363–85.

Woodland, H.R. & Gurdon, J.B. (1968b). The cytoplasmic control of nuclear activity in animal development. *Biol. Rev.*, **43**, 233–67.

Woodland, H.R. & Gurdon, J.B. (1969). RNA synthesis in an amphibian nuclear-transplant hybrid. *Devl. Biol.*, **20**, 89–104.

Yamada, T. (1958). Embryonic induction. In: *A Symposium on the Biochemical Basis of Development* (eds. W.D. McElroy & B. Glass), pp. 217–38. Baltimore: Johns Hopkins Press.

Yamaguchi, N., Kurashige, S. & Mitsuhashi, S. (1973). Immune response in *Xenopus laevis* and immunochemical properties of the serum antibodies. *Immunology*, **24**, 109–18.

Zoond, A. & Rimer, G. (1934). The mechanism of equilibration in *Xenopus laevis*. *Trans. R. Soc. S. Afr.*, **22**, xxiii.

Zwarenstein, H. (1953). A demonstration of the secretion of pepsin by the isolated frog stomach. *Q. Jl. exp. Physiol.*, **38**, 217–23.

Zwarenstein, H. & Bosman, L.P. (1933). The influence of hypophysectomy on the blood sugar and glucose tolerance in *Xenopus laevis*. *Q. Jl. exp. Physiol.*, **22**, 45–8.

Zwarenstein, H. & Duncan, D.G. (1944). Ten years of the *Xenopus* pregnancy test. *Clin. Proc. Cape Town*, **3**, 186–93.

Zwarenstein, H., Sapieka, N. & Shapiro, H.A. (1946). *Xenopus laevis, a bibliography.* Cape Town: The African Bookman.

List of References added in Proof

Bluemink, J.G. & DeLaat, W. (1973). New membrane formation during cytokinesis in normal and cytochalasin B-treated eggs of *Xenopus laevis. J. Cell Biol.*, **59**, 89–108.

Chung, S-H., Gaze, R.M. & Stirling, R.V. (1973). The maturation of toad visual units. *J. Physiol.*, **230**, 57–8*P*.

Dehn, P.F. & Wallace, R.A. (1973). Sequestered and injected vitellogenin: alternative routes of protein processing in *Xenopus* oocytes. *J. Cell Biol.*, **58**, 721–4.

Emilio, M. & Shelton, G. (1974). Gas exchange and its effect on blood gas concentrations in the amphibian, *Xenopus laevis. J. exp. Biol.*, **60**, 567–79.

Gaze, R.M., Chung, S-H. & Keating, M.J. (1972). Development of the retinotectal projection in *Xenopus. Nature (New Biol.) Lond.*, **236**, 133–5.

Grey, R.D., Wolf, D.P. & Hedrick, J.L. (1974). Formation and structure of the fertilization envelope in *Xenopus laevis. Devl. Biol.*, **36**, 44–61.

Ikenishi, K., Kotani, M. & Tanabe, K. (1974). Ultrastructural changes associated with UV irradiation in the 'germinal plasm' of *Xenopus laevis. Devl. Biol.*, **36**, 155–68.

Jurd, R.D. & McLean, N. (1974). Detection of haemoglobin in red cells of *Xenopus laevis* by immunofluorescent double labelling. *J. Microscopy*, **100**, 213–7.

Jurd, R.D. & Stevenson, G.T. (1974). Immunoglobulin classes in *Xenopus laevis. Comp. Biochem. Physiol.*, **48B**, 411–7.

Kalt, M.R. (1972). Ultrastructural observations on the germ line of *Xenopus laevis. Z. Zellforsch.*, **138**, 41–62.

Lázár, G.Y. (1973). The development of the optic tectum in *Xenopus laevis*: a Golgi study. *J. Anat.*, **116**, 347–55.

Roeder, R.D. (1974). Multiple forms of DNA-dependent RNA polymerase in *Xenopus laevis.* Levels of activity during oocyte and embryonic development. *J. biol. Chem.*, **249**, 241–56.

Rosbash, M. & Ford, P.J. (1974). Polyadenylic acid-containing RNA in *Xenopus laevis* oocytes. *J. molec. Biol.*, **85**, 87–102.

Schroeder, T.E. (1973). Cell constriction: contractile role of microfilaments in division and development. *Amer. Zoologist*, **13**, 949–60.

Scott, T.M. (1974). The development of the retinotectal projection in *Xenopus laevis*: an autoradiographic and degeneration study. *J. Embryol. exp. Morph.*, **31**, 409–14.

Tanabe, K. & Kotani, M. (1974). Relationship between the amount of 'germinal plasm' and the number of primordial germ cells in *Xenopus laevis. J. Embryol. exp. Morph.*, **31**, 89–98.

Tymowska, J. & Fischberg, M. (1973). Chromosome complements of the genus *Xenopus. Chromosoma*, **44**, 335–42.

Wall, D.A. & Blackler, A.W. (1974). Enzyme patterns in two species of *Xenopus* and their hybrids. *Devl. Biol.*, **36**, 379–90.

Wyrick, R.E., Nishihara, T. & Hedrick, J.L. (1974). Agglutination of jelly coat and cortical granule components and the block to polyspermy in the Amphibian *Xenopus laevis. Proc. natn. Acad. Sci. U.S.A.*, **71**, 2067–71.

Index

Page numbers in **bold type** include an *illustration* of the item.